"The Unreasonable Effectiveness of Mathematics..."

Eugene Wigner, physicist and mathematician.

Real World Mathematics

Wei Khim Ng
Department of Physics
National University of Singapore

Rajesh R. Parwani
Simplicity Research Institute
Singapore

Simplicity Research Institute, Singapore
www.simplicitysg.net

Real World Mathematics
Published by the Simplicity Research Institute, Singapore
www.simplicitysg.net

Cover design by Wei Khim Ng

For bulk orders, discounts, and permission to use extracts, please contact the Simplicity Research Institute at
enquiry@simplicitysg.net

A CIP record for this book is available from the
National Library Board, Singapore

(Updated Feb 2016)
ISBN: 978-981-09-1163-8

Contents

Preface

As students progress through their high-school mathematics syllabus, they typically find the topics becoming increasingly abstract. Why study logarithms, the periodicity properties of trigonometric functions, or matrix multiplication? Why is it useful to know the solution to a quadratic equation or the derivative of a function?

We believe that the alienation many students feel towards mathematics can be overcome by motivating the various concepts through realistic examples. However, since many such examples involve subjects beyond mathematics, the teachers need to be comfortable handling potential questions the students might have about the context of those questions.

This book hopes to address those concerns. In the first part, from Chapter One till Chapter Eight, we have listed exercises with real-world scenarios. Answers are provided with brief explanatory notes where necessary. However, we do not discuss the mathematical concepts or techniques themselves, such as how to solve a quadratic equation: We assume the reader will be using some other standard text for that introduction, if required. (Nevertheless, for the convenience of users, we have listed the key formulae in an appendix).

In some questions we have suggested investigations and challenges for the interested reader.

Some readers might wish to start from the second part of the book. There we give a longer discussion of key ideas from economics and the physical sciences which provide the background to most of our questions in part one. Included are discussions of mathematical modelling and the scientific method. We have not offered answers to the suggested investigations in part two.

We welcome feedback and questions from users of this book; please email us at **enquiry@simplicitysg.net**. Updates will be posted on the book's web-page:

www.simplicitysg.net/books

A curious reader might wonder why we have not discussed topics such as probability theory or statistics in this book: We feel that those topics are more easily motivated, and covered in many existing texts.

Acknowledgements

Some exercises and notes in this book have been reproduced, with minor modification, from another book by one of the current authors: "Integrated Mathematics for Explorers" by Adeline Ng and Rajesh R. Parwani. We thank the publisher for granting us permission.

Rajesh is grateful to Lynette Soh and Dr. Tan Yingrou for reading parts of the draft manuscript and providing helpful feedback.

Wei Khim thanks his family and friends for their support and encouragement.

"The Unreasonable Effectiveness of Mathematics in the Natural Sciences," was the title of a paper published by Eugene Wigner in 1960 in the Communications in Pure and Applied Mathematics. In that paper he remarks: "The miracle of the appropriateness of the language of mathematics for the formulation of the laws of physics is a wonderful gift which we neither understand nor deserve".

It is a gift we should all learn to use and appreciate.

Wei Khim Ng

Rajesh Parwani

July 2014, Singapore.

Chapter 1

Polynomials and Rational Functions

Among many occurrences, quadratic polynomials appear in chemistry when solving for equilibrium concentrations in a reaction, in physics when studying projectile motion, or as approximations to a curve near its turning point.

Higher degree polynomials also have some real-world applications, as discussed in the exercises. In general, polynomial equations of degree larger than 2 are difficult to solve analytically, but sometimes, if one root can be guessed, they can be reduced by division to lower degree polynomials. Otherwise, approximate solutions may be obtained using numerical methods.

A rational function is a ratio of two polynomials. An example is the formula $v = \dfrac{fu}{u-f}$ which gives the image distance of the image formed by a thin lens of focal length f, when the object distance is u.

1.1 Exercises

1. One publisher's estimated cost $C(x)$, in dollars, to produce x copies of a particular book is

$$C(x) = 10x + 400 \ .$$

The company intends to sell the book at \$30. [*See Sect. 10.3*]

(a) How many books must the company sell to break-even? [Ans: 20]

(b) How many books must the company sell to make a profit of $10000. [Ans: 520]

(c) What is the minimum number of books the company must sell to make a profit of at least 20% over the cost price? [Ans: 27]

2. When a rigid object is immersed in an incompressible fluid, the volume of fluid displaced is the same as the volume of the submerged solid.

   (a) A rectangular block A of dimensions $a \times a \times b$ and a sphere B of radius r are placed in water. One-third of A remains afloat while only one-quarter of B remains afloat. What is the total volume of water displaced? [Ans: $2a^2b/3 + \pi r^3$]

   (b) Discuss how you would determine the density of a small metal statue. [*See Note 1*]

3. The pressure due to a column of water of height h is given by $P = \rho g h$ where ρ is the density of water and g the acceleration due to gravity [*See Sect.11.9*].

   (a) Taking the density of water to be 1000 kg/m^3 and $g = 9.8$ m/s^2, calculate the pressure due to a 10 m water column. [Ans: About $100,000$ N/m^2].

   (b) Normal atmospheric pressure is equivalent to the pressure calculated in part (a). What would the pressure be on an object 10 m below sea-level? [Ans: About 2 atmospheres]

4. Coulomb's law states that the magnitude of the electrical force F between two given point charges varies with their separation R as $F \propto \frac{1}{R^2}$ [*See Sect.12.1.1*]. How far must the charges be separated to half the original force between them? [Ans: $\sqrt{2}R$]

5. When two resistors R_1 and R_2 are connected in parallel in a electrical circuit, their combined resistance R is determined by $\dfrac{1}{R} = \dfrac{1}{R_1} + \dfrac{1}{R_2}$. All resistances are positive. [*See Sect.12.1.3*]

(a) Prove that $R \le R_1$ and $R \le R_2$.

(b) If the value of R_1 is fixed while R_2 is a variable resistor, determine the range of values for R. [Ans: $0 \le R \le R_1$]

(c) Express R as a rational function of R_1 and R_2. [Ans: $R_1R_2/(R_1 + R_2)$]

6. Neo blows across the open end of a cylinder which is closed at the other end. That creates a sound whose frequency is given by $f = v/(4L)$, where v is the speed of sound and L is the length of the cylinder. [*See Note 2*]

   (a) Calculate f for $v = 345$ m/s and $L = 0.2$ m. [Ans: 431.25 Hz]

   (b) Would the frequency in part (a) be audible? [Ans: Probably]

   (c) Experiment making sounds with cylinders of different length, and if possible (if you have a musical ear) try to estimate the frequencies you generate. Compare with the formula's prediction $f \propto 1/L$.

7. In a particular chemical reaction, one molecule of substance A combines with one molecule of substance B to produce one molecule of substance C. If initially there is none of C in the mixture, the eventual concentration, x, of C at equilibrium is determined from the equation $k = \dfrac{x}{(a-x)(b-x)}$ where k is the equilibrium constant, and a, b are the initial concentrations of the reactants. The value of x cannot exceed a nor b.

   (a) If $k = 0.9$, $a = 0.3$ and $b = 0.4$, calculate x. [Ans: 0.069]

   (b) Could solving for the equilibrium concentration in some other chemical reactions lead to higher degree polynomials? [Ans: Yes]

8. A rectangle of sides a and b, with $b/a > 1$, is divided into a square $a \times a$ and a rectangle $a \times (b-a)$ such that $\dfrac{b}{a} = \dfrac{a}{b-a} \equiv \phi$.

   (a) Determine the numerical value of ϕ, known as the "Golden Ratio". [Ans: $(1 + \sqrt{5})/2$]

(b) Study some properties and applications of the Golden Ratio.

9. A sports scientist estimates the maximum height, H m, that any pole-vaulter can reach using the formula $H = \dfrac{v^2}{2g} + 0.9$ where v is the run up speed of the vaulter and $g = 9.8$ m/s^2 is the acceleration due to gravity.

   (a) If a vaulter reaches 6 m, estimate his run up speed. [Ans: 10 m/s]

   (b) If the same vaulter halved his run up speed, how high would he reach? [Ans: 2.18 m]

   (c) Using estimates of the run up speed of actual pole-vaulters, compare the formula's predictions with data on pole-vaulting heights.

10. The A-series of paper sizes start with the largest, $A0$, having an area 1 m^2. If the sides of the (rectangular) $A0$ sheet are denoted by x and y, with x the longer side, then the $A1$ sheet is obtained by dividing the $A0$ sheet in half along its longer side, obtaining dimensions $x/2$ by y, with y now the longest side. If the aspect ratios of the two sizes are equal, that is, $\dfrac{x}{y} = \dfrac{y}{(x/2)}$,

    (a) Determine x and y. [Ans: $x = 2^{1/4}$, $y = 1/x$]

    (b) If succeeding paper sizes ($A2$, $A3$, $A4$...) are defined by dividing each previous size by two along the longer side, determine the dimensions of the $A4$ sheet. [Ans: 297 mm × 210 mm]

    (c) The B-series of paper sizes is obtained by taking the geometric mean of adjacent A-series sizes, so that BN would be the geometric mean of the $A(N-1)$ and AN sizes. What would the size of a $B4$ sheet be? [Ans: 250 mm × 353 mm]

11. Neo is standing on a platform which is at a height y_0 above ground. He throws a ball vertically upwards at time $t = 0$, with

initial speed u. Ignoring air-resistance, the subsequent height of the ball is given by $y(t) = y_0 + ut - \frac{gt^2}{2}$ where g is the acceleration due to gravity and t the time. [*See Note 3*]

(a) Determine the time it would take the ball to reach the ground if $y_0 = 5$ m, $u = 10$ m/s and $g = 10$ m/s^2.
[Ans: $t = 1 + \sqrt{2}$ s]

(b) Determine the maximum height reached by the ball and the time when it reaches that maximum height.
[Ans: $h_{max} = 10$ m at $t = 1$ s]

(c) For how long did the ball stay 2 m or more above ground?
[Ans: $1 + 2\sqrt{10}/5$ s]

(d) If Neo were to throw the ball the same way on the Moon, where the g value is 1/6-th of that on Earth, what would the answers to parts (a, b) and (c) be? [Ans: (*a*) $6 + \sqrt{42}$ s, (*b*) $h_{max} = 35$ m at $t = 6$ s, (*c*) $6(1 + \sqrt{110}/10)$ s]

12. A student attempts to model the population of a bacterial species in a culture by [*See Note 4*]

$$N(t) = N_0 - ct + at^2 .$$

Here t is the time variable, a and c are certain positive constants, and the initial number of bacteria in the culture is N_0. Assuming the model to be approximately correct, with $N_0 = 100$, $c = 5$ and $a = 1$,

(a) How long does it take for the population to double?
[Ans: 12.8 s]

(b) When does the population drop to 94?
[Ans: At $t = 2$ or 3]

(c) Can the population drop to 93? Explain. [Ans: No]

(d) A change in the environment creates the additional term bt^3 on the right-hand-side of the population equation. If $b = 2$,

i. What will be the minimum population in this situation? [Ans: 98]

ii. Use your favourite software to determine when the population will reach 200. [Ans: $t = 3.74$]

13. A scientist uses a high-speed camera to study the trajectory of a ball shot from a tennis ball cannon. He finds that the trajectory is well modelled by $y = ax - bx^2$ with the origin, $(x = 0,\ y = 0)$, at the mouth of the cannon and the ground at $y = -1.500$ m. The ball reaches a maximum height of 51/4 m above ground and hits the ground at $x = 80.46$ m. Obtain answers to the questions below correct to two significant figures. [*See Note 3*]

    (a) Determine the values of a and b (both positive) and sketch the trajectory. [Ans: $a = 0.58,\ b = 0.0074/\text{m}$]

    (b) What are the coordinates (in metres) of the maximum point in the trajectory? [Ans: $x = 39,\ y = 11$]

    (c) At what value of x does the ball reach $y = 0$ again? [Ans: $x = 78$]

    (d) For which range of x values does the trajectory satisfy $y \geq 10$? [Ans: $26 \leq x \leq 52$]

14. The Van der Waals equation for a gas is

$$\left(p + \frac{3}{v^2}\right)\left(v - \frac{1}{3}\right) = \frac{8T}{3},$$

where p, v, T are known as the reduced pressure, volume and temperature. [*See Note 5*]

    (a) Solve the equation for v when $p = T = 1$ (these are called the "critical values"). [Ans: $v = 1$]

    (b) Re-write the Van der Waals equation in partial fraction form, expressing p in terms of v. [Ans: $p = -3/v^2 + (8T/3)(1/(v - 1/3))$]

15. When a spherical tank of radius R is filled with fuel to a height h ($\leq 2R$), the volume of fuel is given by

$$V = \frac{\pi h^2}{3}(3R - h)\ .$$

(a) Verify the correctness of the formula when $h = 2R$.

(b) If $R = 5$ and $V = 36\pi$, find the cubic equation that determines h and find one of its roots by trial. [Ans: $h = 3$]

(c) Find all the positive roots of the cubic equation in part(b) and determine which of them are physically permissible. [Ans: $h = 3,\ 6(1+\sqrt{2})$. But $h \le 2R$ implies only $h = 3$ is permissible]

(d) Challenge: Use calculus to prove the formula relating V to h.

16. You are instructed to construct an open top box from a square, $a \times a$, piece of cardboard: First cut out squares of dimensions x by x at each corner of the cardboard, then fold the cardboard to form a box of dimensions x by $(a-2x)$ by $(a-2x)$.

(a) Find the values of x which make the volume of the box a minimum and explain the physical situation those values correspond to. [Ans: $V = 0$ when $x = 0$ (base area but no height) or $x = a/2$ (height but no base area)]

(b) Hence, or otherwise, explain why for some value of x the volume of the box would be a maximum. Determine that value of x. [Ans: $x = a/6$]

(c) If $a = 2$, estimate the maximum volume of the box using an appropriate plot. [Ans: $V_{max} = 16/27$ at $x = 1/3$]

(d) Solve for part(c) using calculus instead of a plot.

17. Consider motion along a straight line. An object is observed from two frames of reference, as shown in Fig.(1.1). Its velocity relative to the stationary person (S) is v. The second observer (S′), who has velocity u relative to the stationary observer, measures the object's velocity to be v'. In Einstein's Special Theory of Relativity, the relative velocity formula is

$$v' = \frac{v-u}{1-\frac{uv}{c^2}}\,. \tag{1.1}$$

The constant c is the speed of light. [*See Note 6*]

(a) Invert the relative velocity formula to obtain v in terms of the other variables. [Ans: $v = \dfrac{v' + u}{1 + uv'/c^2}$]

(b) Show that $v = c$ implies $v' = c$; that is, the speed of light has the same value in the two frames of reference.

(c) Find all possible values of v for which $v = v'$ even when $u \neq 0$. [Ans: $v = \pm c$]

(d) Challenge: Show that for $v/c \ll 1$ and $u/c \ll 1$, the relative velocity formula reduces to what you would expect in Galilean-Newtonian physics, $v' \approx v - u$. [*See Sect.11.2*]

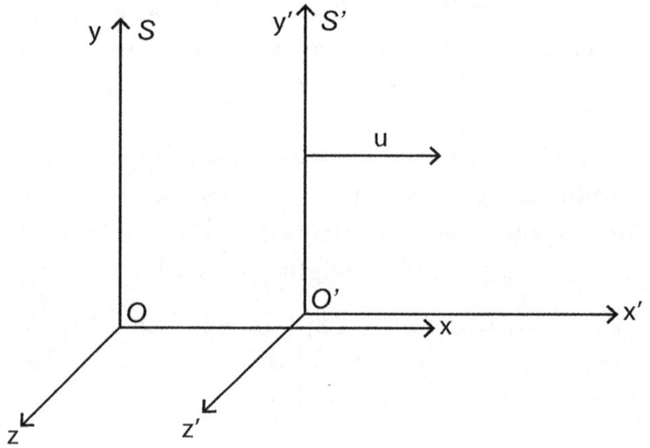

Figure 1.1: Two inertial frames in relative motion

18. The thin lens formula in optics relates the object distance, u, to the image distance, v, where both u, v are measured relative to the centre of the lens. For a converging lens the formula is $\dfrac{1}{u} + \dfrac{1}{v} = \dfrac{1}{f}$ where $f > 0$ is the constant focal length of the lens.

    (a) Find a partial fraction expression for v in terms of u. [Ans: $v = f + \frac{f^2}{u-f}$]

    (b) If the object distance ($u > 0$) is varied, determine the range of values for the image distance. [Ans: $-\infty \leq v \leq \infty$]

19. An engineer uses experimental data to determine a particular function $f(x)$ she requires for her engineering design. She finds that near one turning point the function may be approximated by $2x^2 - 3x + 1$.

    (a) Determine the value of the function at that turning point and the nature of the turning point. [Ans: Local minimum at $x = 3/4$ with $f(3/4) = -1/8$]

    (b) Could the function $f(x)$ have another turning point? [Ans: Possibly. $f(x)$ is an approximation to the actual (unknown) function]

20. For Q13 in Chap.7, the amount of cement in the cone at time t is given by $V(t) = 5t^2 - \frac{t^3}{3} + 10$. Let $t = T$ be the time when there is no more cement in the cone.

    (a) Show by direct substitution, and using a sketch of $V(t)$, that $15 < T < 15.5$.

    (b) Writing $T = 15 + x$, and setting $V(T) = 0$, find an approximate linear equation for x. (That is, ignore quadratic and higher order terms in x. [Ans: $10 - 75x = 0$]

    (c) Hence deduce an approximate value for T. [Ans: $T = 15.1333$]

    (d) Repeat the step in part (b), but now keep terms quadratic in x to find another approximation for T. [Ans: $T = 15.13104$]

    (e) Use software to determine T accurately. Compare with your approximations above. [Ans: $T = 15.13103$]

1.2 Notes

1. Q2a: This is related to the famous story of Archimedes.

2. Q6: The speed of sound depends on the temperature and other properties of the medium. The wave formed in the cylinder in this question is an example of a standing wave, with zero amplitude at the closed end and maximum amplitude at the open end; see also Q3 of Chap.4.

3. Q(11, 13): These projectile motion equations can be derived from Newton's laws (Sect.11.3), whereby the accelerations in the y and x directions are given by $\frac{dv_y}{dt} = -g$ and $\frac{dv_x}{dt} = 0$ if one ignores air-resistance. See also Q12 from Chap.4. In these problems g is the constant acceleration due to gravity– however this assumes that one is not too far above the surface of the Earth as g actually varies (Sect.11.6). Air-resistance can be taken into account, see the exercises in the next chapter.

4. Q12: Generally, bacteria replicate via binary fission, that is, one cell splits into two. This leads to a typical exponential growth in population if there are no constraints (see Eq.2.2). In this exercise, we are assuming that the culture is exposed to stresses which cause cells to reproduce at different rates, so that the population growth is not exponential during that time interval. See also Sect.9.4.2.

5. Q14: The Van der Waals equation gives a better description of gases at low temperatures or high pressures than the popular ideal gas law $PV \propto T$ (Sect.11.8). It takes into account the finite size of molecules and their interaction with each other.

6. Q17: An object's velocity is a relative quantity, depending on the reference frame (Sect.11.2). For example, suppose a stationary person A and a moving person B both observe a flying aeroplane. Relative to themselves, A and B will each measure a different velocity for the aeroplane.
   The formula in the question relates those two observations by A and B in the context of Einstein's Theory of Special Relativity. See also Q4 from Chap.6.

   Typically, the predictions of Einstein's theory differ significantly from those of Galileo and Newton only at high velocities (Sect.13.1).

Chapter 2

Exponents and Logarithms

One practical application of logarithms is in compressing a large range of data values; for example, if a variable x takes values from 10^{-7} to 10^7 then $\log_{10} x$ ranges only from -7 to 7. As such, logarithms appear when quoting magnitudes of earthquakes, the intensity of sound, or the acidity of solutions, among other applications.

The logarithm is inversely related to the exponential function through

$$y = b^x \Leftrightarrow x = \log_b y \ . \tag{2.1}$$

Exponential functions occur in many mathematical models of natural phenomena. For example, if

$$\frac{dx}{dt} = kx \ , \tag{2.2}$$

then

$$x \propto e^{kt} \ , \tag{2.3}$$

where $e = 2.718...$ is the base of natural logarithms. For $k > 0$, Eq.(2.2) represents a growth process, such as for biological cells (for limited times), while for $k < 0$ the expression can be used to model decay processes such as in radioactivity.

2.1 Exercises

1. The speed of light in vacuum is 299792458 m/s. Express it in standard (scientific) notation. [Ans: 2.99792458×10^8 m/s.]

2. A laser beam sent from Earth and reflected from the Moon returns in 2.6 s. Taking the speed of light to be 3.0×10^5 km/s, calculate how far the Moon is from Earth.
[*See Note 1*] [Ans: 3.9×10^5 km]

3. In astronomy, the unit of distance "light-year" is defined as the distance light would travel in one year. [Note: In this question, all speed and time observations are relative to Earth].

   (a) How many km's is a light-year? [Ans: 9.46×10^{12} km]

   (b) If star A is 6 light-years from us, how many years will it take a spaceship from Earth travelling at (the futuristic speed) $v = 0.2c$ to reach it? (Hint: it might be useful to write one light-year as $c.y$ reminding you that it is the speed of light c multiplied by one year.) [Ans: 30 yrs]

   (c) Another spaceship makes the same journey to star A in 18 years. What was its speed in terms of c? [Ans: $c/3$]

   (d) If the spaceship in part(c) continues its journey beyond star A for another 12 years, what would its total distance from Earth be? [Ans: 10 light years]

4. 18 grammes of water contain about 6.02×10^{23} molecules.
[*See Sect.11.1*]

   (a) How many molecules are there in 1 litre of water?
   [Ans: 3.34×10^{25}]

   (b) How much do 1000 molecules of water weigh?
   [Ans: 3×10^{-20} g]

5. The period, T, of a simple pendulum is the time taken for one complete cycle. A student carefully studies how the period varies with the length, l, of the pendulum by plotting $\log T$ against $\log l$. The data fit to a straight line of slope 1/2.
[*See Note 2*]

   (a) How does T vary with l? [Ans: $T \propto \sqrt{l}$]

   (b) If the length is doubled, what happens to the period?
   [Ans: $T \to \sqrt{2}T$]

6. In Einstein's Special Theory of Relativity, the energy of a particle moving at speed v is given by $E = \dfrac{mc^2}{\sqrt{1 - v^2/c^2}}$ where c is the speed of light and m the mass of the particle. [*See Sect. 13.1*]

   (a) What would the energy of the particle be when $v = 0$? [Ans: mc^2]

   (b) If E and m are real and finite, what does Einstein's expression imply for the maximum speed of a particle? [Ans: c]

   (c) Challenge: Find the first two terms in the expansion of E for $v/c \ll 1$.

7. The speed of a tsunami in water of depth d is given by $v = \sqrt{gd}$ where $g = 9.8$ ms^{-2} is the acceleration due gravity. [*See Note 3*]

   (a) Calculate the speed of a tsunami when $d = 2$ km. [Ans: 140 m/s]

   (b) If the speed of a tsunami decreases by a factor of 3 as it moves towards shore, by what factor has the water depth decreased? [Ans: 1/9]

8. Yeast in a certain medium grows to double its mass every 40 hours. The mass of Yeast at time t hours is represented by $N(t)$. [*See Note 4*]

   (a) Find an expression for $N(t)$ when $t = 40$, 80 and 120 hours. Express your answer in terms of $N(0)$. [Ans: $2N(0)$, $4N(0)$, $8N(0)$.]

   (b) Hence explain why $N(t) = 2^{t/40}N(0)$.

   (c) If $N(0) = 5$g, how much yeast is there after 20 hours? [Ans: 7.07 g]

   (d) If $N(0) = 10$g, how long does it take the yeast to reach 30g? [Ans: 63.4 hrs]

9. Moore's Law is an empirical observation in microelectronics. It states that the number of transistors on integrated circuits doubles every two years. [*See Note 5*]

(a) Express the law in mathematical form.
[Ans: $N(t) = N(0)2^{t/2}$ with t in years]

(b) When did the number of transistors on integrated circuits reach twenty times the value that was in the first year?
[Ans: $t = 8.64$]

10. The author of a mathematics book models the sales volume $S(t)$ of his previous best seller by the function $S(t) = C(1 - e^{-kt})$, where C and k are constants and t is the time in months since the release of the book.

(a) Sketch $S(t)$.

(b) If he sold about 20,000 copies, and if the current monthly sales of the book are negligible, estimate C.
[Ans: $C = 20,000$]

(c) If he reached half of the total sales just in the first six months, estimate k. [Ans: $k = (\ln 2)/6$]

(d) What was his total sales by the end of the first year?
[Ans: 15,000 books]

11. When radioactive atoms decay, they change into atoms of a different type. If the initial number of radioactive atoms is N_0, then the number that remain unchanged at time t is given by the equation $N(t) = N_0 e^{-\lambda t}$ where λ, called the "decay constant", depends on the element. [*See Sect. 13.3*]

(a) Determine, in terms of λ, the time taken for the number of radioactive atoms of a specific type to drop to half its starting value. (This value, denoted by $t_{1/2}$, is called the half-life of that element.) [Ans: $\ln 2/\lambda$]

(b) A scientist estimates that the amount of (radioactive) Carbon-14 remaining in a sample is 1/5 of its original value. Given that the half-life of Carbon-14 is 5730 years, estimate the age of the sample. [Ans: 13305 yrs]

(c) The scientist finds another radioisotope in a different sample. He estimates the second sample to be 75,000 years old and the radioisotope present to be at 12% of its original

value. What is the half-life of the radioisotope?
[Ans: $\approx 24,500$ yrs]

(d) The decay law $N(t) = N_0 e^{-\lambda t}$ is written in base e. Rewrite it in base 1/2. [Ans: $N(t) = N_0 \left(\frac{1}{2}\right)^{\frac{t}{t_{1/2}}}$]

12. A real estate investment company projects a revenue growth of 5% per annum for the next few years. If its current annual revenue is \$100 million,

(a) What is the projected annual revenue three years from now? [Ans: \$115.76 million]

(b) If the projected growth continues, how long would it take for the annual revenue to reach \$200 million?
[Ans: 14.2 yrs]

13. The pH value of a solution indicates its hydrogen ion concentration, denoted by $[H^+]$ in moles per litre, through the formula

$$pH = -\log_{10}[H^+] \ .$$

Neutral solutions have a pH of 7, acidic solutions have pH values below 7, while base solutions have a pH above 7.

(a) Determine the hydrogen ion concentration of pure water, a neutral solution. [Ans: 10^{-7} moles/litre]

(b) Do acidic solutions have a larger, or smaller, hydrogen ion concentration than neutral solutions? [Ans: Larger]

(c) If a solution has a hydrogen ion concentration of 5×10^{-7} moles/litre, determine its pH value. Is the solution acidic or alkaline (basic)? [Ans: 6.3, acidic]

(d) Is the rainfall in your region acidic or basic? What are the implications?

(e) Is the soap you use acidic or basic? Which is better for you?

14. A convenient way to compare the intensity I of a sound to a reference intensity I_0 is through the decibel (dB) scale defined by $L = 10 \log_{10} \left(\frac{I}{I_0}\right)$.

(a) If $L = 30$ dB, what is the ratio of the two sound intensities? [Ans: 1000]

(b) If $I = 4I_0$, what is the dB value? [Ans: 6]

(c) What dB levels are considered harmful to the human ear?

15. On the Richter scale, the magnitude R of an earthquake is determined by

$$R(I) = \log \frac{I}{I_0} ,$$

where I is the measured amplitude and I_0 a reference value. The logarithms are in base 10. [*See Note 6*]

(a) An earthquake of magnitude 8.3 struck the Sea of Okhotsk (Russia) in 2013. What was the measured amplitude? Express your answer in terms of I_0. [Ans: $1.995 \times 10^8 I_0$]

(b) Also in 2013, an earthquake in the South Atlantic ocean registered an amplitude of $I = 10^7 I_0$. What was its magnitude on the Richter scale? [Ans: 7]

(c) The energy U released by an earthquake is related to its Richter magnitude by

$$R(U) = \frac{2}{3} \log \frac{U}{U_0} ,$$

where U_0 is a reference value. Find the ratio of the energies, U_1/U_2, released by the two earthquakes in parts (a) and (b) above. [Ans: 89].

16. Rydberg's formula for hydrogen describes the wavelength, λ, of radiation emitted in transitions of the atom. With integers $n_2 > n_1 \geq 1$, the formula is $\frac{1}{\lambda} = R\left(\frac{1}{n_1^2} - \frac{1}{n_2^2}\right)$ where $R = 1.097 \times 10^7\ m^{-1}$ is Rydberg's constant. [*See Note 7*]

(a) Calculate the shortest wavelength that a hydrogen atom can emit in the $n_1 = 1$ (Lyman) series.
[Ans: 9.12×10^{-8} m]

(b) Calculate the longest wavelength that a hydrogen atom can emit in the $n_1 = 2$ (Balmer) series.
[Ans: 6.56×10^{-7} m]

(c) Find a simplified expression for λ in terms of n_1 and n_2.
[Ans: $\lambda = \dfrac{n_1^2 n_2^2}{R(n_2^2 - n_1^2)}$]

17. The curve formed by a freely hanging cable supported only at its ends is called a catenary. Its equation is $y = \frac{a}{2}\left(e^{x/a} + e^{-x/a}\right)$.
[*See Note 8*]

(a) If the lowest point of the cable reaches 100 m below its support level, and if $a = 10$ m, determine the coordinates, in metres, of the support point.
[Ans: $x = \pm 30.89$, $y = 110$]

(b) Challenge: Show that a catenary may be approximated by a parabola near its bottom.

18. The density, ρ, of the nucleus of an atom is roughly constant, independent of the type of atom. Also, the mass of a nucleus is proportional to the number of constituent nucleons A. Assuming the nucleus to be spherical, express the nuclear radius R in terms of the number of nucleons, A. Denote the mass of each nucleon by m and ignore the gaps between nucleons.
[*See Note 9*] [Ans: $R = (3m/4\pi\rho)^{1/3} A^{1/3}$]

19. A bank pays 2% compound interest per month on its fixed deposits. A lady deposits \$10,000. Do your calculations below using two methods: The binomial theorem and directly, using a calculator. [*See Note 10*]

(a) How much would she obtain when she makes a withdrawal six months later? [Ans:\$11,261.62]

(b) If she decides not to make any withdrawal at the end of 6 months, but wishes to keep her deposit until the total amount reaches \$12,000, how many months would she need to wait? [Ans: 10 (rounded)]

20. What effective annual interest rate corresponds to a 5% annual interest compounded monthly? [*See Sect.10.1*] [Ans: 5.12%]

21. An investor wishes to have \$15,000 accumulated at the end of two years from his savings in a fixed deposit account which pays 3% per annum compounded each quarter. How much should he deposit initially to achieve his goal? [Ans: \$14,130]

22. An investor hopes to have \$20,000 accumulated at the end of three years by placing his savings in a fixed deposit account which pays $R\%$ per annum compounded each quarter. If he has \$12,000 initially, what value of R would allow him to achieve his goal? [Ans: $R = 17.4$]

23. Newton's law of cooling describes how an object cools,

$$\frac{dT}{dt} = -k(T - T_a) \ ,$$

where $T(t)$ is the varying temperature of the object, t is time, T_a is the constant ambient temperature, and $k > 0$ is a constant.

(a) Defining $y = T(t) - T_a$, show that $\dfrac{dy}{dt} = -ky$.

(b) Show that $y = Ae^{-kt}$ is a solution of the last equation.

(c) Hence obtain an explicit formula for $T(t)$ if $T(t = 0) \equiv T_i$. [Ans: $T(t) = T_a + (T_i - T_a)\, e^{-kt}$]

(d) Verify that your solution in the last step gives the expected result for $t = 0$ and also as $t \to \infty$.

(e) If $k = 0.06$ per minute, determine how long it would take a body at 100°C to cool to 50°C if $T_a = 25$°C. [Ans: 18.31 mins]

(f) What would the temperature of the body in part (d) be 15 minutes after it reaches 50°C? [Ans: 35.2°C]

24. In addition to gravitational acceleration, a particle experiences air-resistance as it falls along its vertical trajectory: Its net acceleration is given by

$$\frac{dv}{dt} = g - bv \tag{2.4}$$

where $v(t)$ is the instantaneous velocity and b, g are positive constants (taking the downward direction as positive). *[See Note 11]*

(a) Show that $v(t) = g(1 - e^{-bt})/b$ satisfies Eq.(2.4) and the initial condition $v(0) = 0$.

(b) Use the last result to determine the constant velocity, known as "terminal velocity", which the particle will reach if it continues to fall for long time. [Ans: g/b]

(c) Show that the terminal velocity may be obtained directly from Eq.(2.4) by setting $dv/dt = 0$.

(d) Challenge: Obtain the result in part(a) by integration of the acceleration equation.

(e) Discuss one application where air-resistance and a low terminal velocity are essential.

25. Suppose that the density of a spherical object varies with radial distance, $r > 0$, as

$$\rho(r) = \frac{\rho_0}{1 + e^{\frac{r-R}{a}}}$$

where a and R are some positive constants. [*See Note 12*]

(a) Show that when $r = R$, $\rho = \rho_0/2$.

(b) Sketch the curve.

(c) Let x be the distance in which the density drops from 90% to 10% of ρ_0. Find an expression for a in terms of x. [Ans: $a = \frac{x}{2\ln 9}$]

(d) If x is 2.4 fm, what is a? [Ans: 0.546 fm]

2.2 Notes

1. Q(1, 2): The speed of light depends on the medium. The usual quoted value is for vacuum, which differs slightly from that in air (Sect.12.2.2). During the Apollo missions, reflectors were actually placed on the moon!

2. Q5: A simple pendulum consists of a small weight supported by a very light string. The string is attached to a pivot about which it can swing. If the amplitude of oscillations is small, the period of oscillation is essentially independent of the amplitude and is described as "simple harmonic motion" (Sect.9.4.3). This

means that the period remains roughly constant even as air-resistance damps the oscillations.

3. Q7: The general formula for water waves is complicated, but since tsunami's have long wavelengths, compared to the water depth, the simple expression used in this problem is valid. The energy carried by a tsunami is dependent on its speed and height, so as the speed decreases near shore the wave height increases dramatically, conserving energy.

4. Q8: Yeast is a type of fungi that breaks down carbohydrates into carbon dioxide and alcohols. This process is called fermentation. Yeast is used in baking, production of alcoholic beverages, and probiotics.

5. Q9: An integrated circuit (IC) consists of electric circuits etched on a tiny piece of material such as silicon. The transistor is one of the main components of an IC. IC's have become ubiquitous in electronic equipment.

6. Q15: The amplitude I is measured on a seismograph located some distance from the quake's epicentre. Beyond Richter, there are other scales, some subjective, which summarise the destructive power of an earthquake.

7. Q16: The Rydberg formula summarises the discrete emission spectrum that was empirically observed. The formula was later derived using quantum theory (Sect.13.2).

8. Q17: If the cable supports a load, such as in a bridge, the shape would change, becoming closer to a parabola. Also, near its lowest point, a catenary may be approximated by a parabola as is seen through a Taylor expansion of the function.

9. Q18: An atomic nucleus consists of protons and neutrons, which have almost the same mass and are collectively called nucleons (Sect.13.3).

10. Q19: Compounding at that rate means that at the end of each month the total amount would be 1.03 times the amount at the start of that month (Sect.10.1).

11. Q24: The air-resistance term, proportional to $-v$, is only a convenient approximation, valid for small objects at low velocity. For larger objects and higher velocities the resistance would be proportional to $-v^2$ (Sect.11.10). Parachutes increase air drag intentionally, thus allowing skydivers to reach a low terminal velocity and reach the ground safely.

12. Q25: This density profile is typical of an atomic nucleus. The distance in which the density drops from 90% to 10% is called the skin thickness. The unit of measure "Fermi" is often used nuclear physics, 1 fm $= 10^{-15}$ m.

Chapter 3

Simultaneous Equations and Matrices

Many real-world problems involve multiple variables that need to satisfy various constraints.

In simple cases, two simultaneous equations may be reduced to one equation by eliminating one common variable between them. However even when this is possible, the resulting equation might be too difficult to solve analytically, as in the example of the two equations $y = 2e^{-x}$ and $y - x - 1 = 0$. In such cases, one approach would be to find an approximate solution using graphical methods.

If the simultaneous equations are all linear, a systematic procedure to solve them exactly uses matrices.

When there is a combination of equations and inequalities, the boundaries of the solution set are typically first determined by treating all of them as equalities, and then imposing the inequalities.

3.1 Exercises

1. On the Celsius temperature scale, the freezing point of water is 0°C while the boiling point is 100°C. On the Fahrenheit scale, commonly used in the USA, those two points have the respective values 32°F and 212°F. The two scales are linearly related, $F = aC + b$, where we have denoted the numerical values on the two scales by F and C.

(a) Determine the values of the constants a, b.
[Ans: $a = 9/5$, $b = 32$]

(b) Express C in terms of F. [Ans: $C = 5(F - 32)/9$]

(c) Is there any temperature for which the Celsius and Fahrenheit scales give the same numerical value? [Ans: $-40°$]

(d) Temperatures T on the absolute scale are measured in Kelvins and are related to the Celsius scale by
$T = C + 273.15$. [*See Sect.11.8*]

i. What Fahrenheit scale temperature F corresponds to the absolute temperature T?
[Ans: $F = 1.8T - 459.67$]

ii. The lowest temperature is absolute zero, $T = 0$. What is the corresponding value on the Celsius and Fahrenheit scales? [Ans: $-273.15°$C, $-459.67°$F]

2. You need to carry 12 bags of apples and 6 bags of grapefruit from the car to a picnic spot. Each bag of apples weighs 2kg while each bag of grapefruit weighs 5kg. You can carry a maximum of 8kg per trip. What is the minimum number of trips you need to make? (The bags are sealed and should not be opened).
[Ans: 8]

3. A car can reach a maximum speed of 80 km/hr. Its fuel consumption (in some units) when moving at v km/hr is $C = (v/20)^3$ units per hour. The car must be driven from one town to another, a distance of 300 km, in less than 6 hours, but the total fuel consumption must not exceed 150 units. Assuming, for simplicity, that the car travels at constant speed, determine the range of values for v which achieve the objective.
[Ans: $50 < v < 63.25$]

4. The force of gravity on an object of mass m close to the surface of the earth is given by Newton's Second Law, $F = mg$, where $g = 9.8$ m/s^2 is the acceleration due to gravity. The same force may be determined from Newton's law of gravity, $F = \frac{GMm}{R^2}$. By equating the two expressions, obtain a value for the mass of the earth M. Use the value $G = 6.67 \times 10^{-11}$ m^3 kg^{-1} s^{-2} for

the gravitational constant and $R = 6370$ km for the radius of the Earth. [*See Sect.11.6 and Q5 from Chap.4*]
[Ans: $M = 5.96 \times 10^{24}$ kg]

5. The owner of a home business produces hand-made dolls which she then sells through her website. The supply and demand functions for her goods are [*See Sect.10.2*]

$$Q_s = 2P - 10\,, \tag{3.1}$$
$$Q_d = 25 - 2P\,, \tag{3.2}$$

where P is the price, Q_s the quantity of goods in supply, and Q_d the quantity of goods for which there is a demand.

   (a) Invert each of the equations to get P as a function of Q. [Ans: $P = 5 + Q_s/2$; $P = 25/2 - Q_d/2$]
   (b) Plot each of the functions in the last part on the same graph (P vs Q).
   (c) Find the equilibrium point where the supply of goods meets demand. What is the value of the equilibrium price and quantity? [Ans: $P = 35/4,\; Q = 15/2$]
   (d) If the seller sets her selling price above its equilibrium value, will there be an excess or deficit in the supply of goods? [Ans: Quantity in supply will exceed demand]

6. Consider motion along one dimension, as described by two frames of reference, one moving relative to the other, as in Fig.(1.1). Spatial coordinates in the stationary frame are labelled by x and t is the time, while the moving frame uses spatial coordinates x' and time t'. The moving frame is moving along the positive x direction with speed u with respect to the stationary frame. In Galilean-Newtonian physics, the coordinates in the two frames are related by [*See Sect.11.2*]

$$t' = t\;,$$

$$x' = (x - ut)\;.$$

Invert the relations to find x and t explicitly in terms of the primed quantities. [Ans: $x = x' + ut'$]

7. In the Special theory of Relativity, the corresponding relations for the situation in the previous question would be the Lorentz transformations described by [*Sect.13.1*]

$$t' = \gamma\left(t - \frac{u}{c^2}x\right) \ ,$$

$$x' = \gamma\,(x - ut) \ ,$$

where c is the speed of light and $\gamma = 1/\sqrt{1-u^2/c^2}$.

   (a) Re-write the Lorentz transformation in matrix form.

   $$\text{[Ans:} \quad \begin{pmatrix} t' \\ x' \end{pmatrix} = \begin{pmatrix} \gamma & -\gamma u/c^2 \\ -\gamma u & \gamma \end{pmatrix} \begin{pmatrix} t \\ x \end{pmatrix} \quad]$$

   (b) Obtain the inverse Lorentz transformation, that is, write the unprimed coordinates in terms of the primed coordinates. Do this using matrices and then check using direct manipulation of the equations.

   $$\begin{aligned} \text{[Ans: } t &= \gamma\left(t' + \frac{u}{c^2}x'\right) \ , \\ x &= \gamma\,(x' + ut') \ . \text{]} \end{aligned} \qquad (3.3)$$

   (c) Challenge: Show that for $u/c \ll 1$ the Lorentz transformation reduces to the Galilean-Newtonian transformation of the previous exercise.

8. A falling object of mass m experiences the force of gravity, $F = mg$, and the opposing force of air-resistance which we will approximate by $F = kAv^2$. k is a constant, A is the surface area of the object perpendicular to its velocity v, and g is the acceleration due to gravity. When the two forces balance, the object reaches a constant "terminal velocity". Determine an explicit expression for the terminal velocity V_T. [*See Sect.11.10*] [Ans: $V_T = \sqrt{mg/kA}$]

9. You need to go on foot from town A to town B which is 42 km away. You are able to run at 6 km/hr for a maximum of 3 hours. You can walk at 3 km/hr. Your ability to move on foot

is limited by a maximum travel time of 13 hours. In addition, you need to take a rest break of 20 minutes after each hour of running (If you stop running before one continuous hour, you will still need a 20 minutes rest break).

(a) What strategy should you use to minimise your travel time? [Ans: Walk for 8 hours, then run for 3 hours, taking two rest breaks before town B and a final rest break at town B]

(b) What would your minimum travel time be? [Ans: 11 hours 40 mins]

10. Archimedes' Principle states that the buoyancy (upward) force, F_u, acting on an immersed (partially or fully) object is equal to the weight of the fluid displaced. If ρ_0 is the density of the fluid, V_0 the displaced volume and g the acceleration due to gravity, then $F_u = \rho_0 V_0 g$. [*See Sect.11.7*]

(a) The downward force on the object is equal to its weight $W = \rho V g$, where V is the volume of the object and ρ its density. By balancing the forces on the object, find an expression for the density of the floating object. [Ans: $\rho = \rho_0 V_0 / V$]

(b) Determine the densities of blocks A and B in Q(2a) of Chap.1 as a fraction of the fluid density. [Ans: Cube density is $2\rho_0/3$, Sphere density is $3\rho_0/4$]

11. A company director obtained emergency loans from three banks to cover his operating expenses, the amount summing to \$1,000,000. The annual interest rates on the loans were 6%, 8% and 10% respectively. The total annual interest payable on the loans was \$70,000.

(a) What amount was borrowed at each rate? Express your answer in terms of \$ x, the loan at 6% [Ans: x at 6%, 1,500,000-$2x$ at 8%, x-500,000 at 10%]

(b) What is the possible range of values for x? [Ans: $500,000 \leq x \leq 750,000$]

(c) It was later revealed that the director borrowed $200,000 more at 8% than at 10%. How much did he borrow at each interest rate?
[Ans: $600,000$ at 6%, $300,000$ at 8%, $100,000$ at 10%.]

12. The 210 students of a school are teamed up into groups of either 3 or 5. The total number of teams formed is N. Let x be the number of teams with 3 members and y the number of teams with 5 members.

(a) If $N = 46$, determine x and y. Solve the problem using two different methods. [Ans: $x = 10,\ y = 36$].

(b) Find the value of N for which x is as close as possible to y. In this case, what are the values of x and y?
[Ans: $N = 52,\ x = 25,\ y = 27$]

13. A finance company manager decides to reward his 245 employees with a trip to the beach. He also decides to support two of his clients, which are transportation providers, by hiring buses from each of them. The buses from provider X can hold 40 passengers each while those from provider Y can hold 30 passengers each. The manager wishes to spend the same amount of money on both providers. If each bus from provider X costs three times as much to rent as a bus from provider Y, what is the minimum number of buses he should hire from each provider?
[Ans: $x = 2,\ y = 6$]

14. Five participants, labelled a, b, c, d, e make it to the finals of a singing contest. Each participant is scored by three judges, J, K and L, on two criteria: Vocals V and originality G. The tables below show the scores given by the judges.

Scores by Judge J

	a	b	c	d	e
V	5	1	3	2	4
G	4	3	5	2	5

Scores by Judge K

	a	b	c	d	e
V	4	2	3	3	3
G	2	3	4	3	5

Scores by Judge L

	a	b	c	d	e
V	3	4	2	4	3
G	1	3	3	4	4

(a) Construct three 2×5 matrices, J, K and L, to represent the scores of the three judges. [Ans: The matrices are essentially as displayed in the tables]

(b) The total raw score for each criteria is the sum of each judges' score for that criteria. Explain how you would construct a matrix T from J, K and L to represent the total raw score for each criteria for the participants. Determine the matrix T explicitly.

$$[\text{ Ans : } T = J + K + L = \begin{pmatrix} 12 & 7 & 8 & 9 & 10 \\ 7 & 9 & 12 & 9 & 14 \end{pmatrix}]$$

(c) The final score for each participant is a weighted average of their total raw scores for each criteria: The vocal factor by 0.7 and the originality factor by 0.3. Construct a 1×2 matrix to represent the "weight" matrix W which when multiplied to T would give the final score F.
[Ans: $W = (0.7,\ 0.3)$]

(d) Determine the matrix F explicitly.
[Ans: $F = (10.5,\ 7.6,\ 9.2,\ 9,\ 11.2)$]

(e) The winner is the participant with the highest final score. Who was the winner? [Ans: Participant e]

15. A company sells backpacks of different sizes and styles. The size x is measured in the company's unit of area, "UA".

(a) The manager prices the backpacks, in dollars, using the formula

$$C(x) = A + Bx \ ,$$

where A, B are constants. The prices of a 10 UA and 20 UA backpack are \$70 and \$120 respectively. What would be the price of a 25 UA backpack? [Ans: \$145]

(b) The price of a water-proofed version of the backpacks is given, in dollars, by

$$S(x) = E + Fx + Dx^2 \ ,$$

where D, E, F are constants. If the prices for 10 UA, 15 UA and 20 UA water-proofed backpacks are \$80, \$111.25 and \$145 respectively, determine the price of a 25 UA water-proofed backpack. [Ans: \$181.25]

(c) The price of a custom made backpack is given, in dollars, by

$$P(x) = 50 + 5x + 0.05x^2 \ .$$

i. What is the area of a custom made backpack if its price is \$117.2? [Ans: 12 UA]

ii. Is there any range of prices for which this model is not available? [Ans: $P < 50$]

16. Consider the food-web illustrated in the figure below. An arrow head pointing at one species indicates that it consumes the species where the arrow starts from. [*See Note 1*]

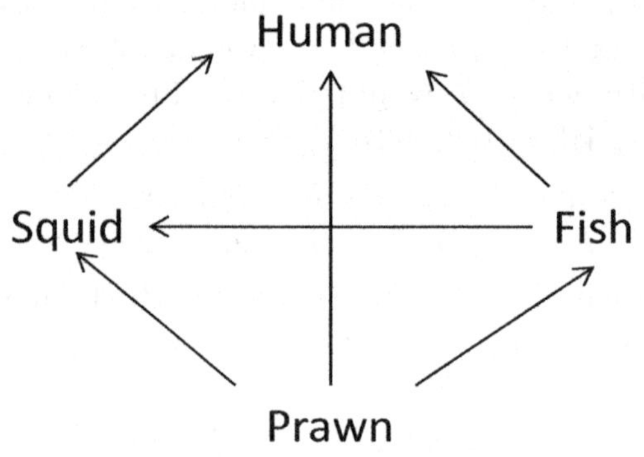

(a) Construct a matrix A in which each row and column represents a species in the food-web. Use the number 1 in a position (m, n) to indicate that a species in row m feeds on the species in column n, otherwise use 0. (Order the rows as "human, squid, fish and prawn", and similarly for the columns). [Ans: $A = \begin{pmatrix} 0 & 1 & 1 & 1 \\ 0 & 0 & 1 & 1 \\ 0 & 0 & 0 & 1 \\ 0 & 0 & 0 & 0 \end{pmatrix}$]

(b) Evaluate A^2. [Ans: $A^2 = \begin{pmatrix} 0 & 0 & 1 & 2 \\ 0 & 0 & 0 & 1 \\ 0 & 0 & 0 & 0 \\ 0 & 0 & 0 & 0 \end{pmatrix}$]

(c) Evaluate A^3. [Ans: $A^3 = \begin{pmatrix} 0 & 0 & 0 & 1 \\ 0 & 0 & 0 & 0 \\ 0 & 0 & 0 & 0 \\ 0 & 0 & 0 & 0 \end{pmatrix}$]

(d) What do the matrices A^2 and A^3 represent?
[Ans: A^2 represents the number of indirect ways one species feeds on another, with one intermediary. For example prawn is "eaten" by humans through fish and squid. A^3 has a similar interpretation.]

3.2 Notes

1. Q(16): Similar uses of matrices appear in investigations of other networks, such as gene networks, chemical networks, friendship networks, and communication networks. The study of complex networks is an important field with many applications, for example in understanding the spread and effective methods of containment of epidemics.

Chapter 4

Geometry and Trigonometry

Some major applications of trigonometry involve the measurement of angles and lengths, for example in navigation and surveying.

However, what makes trigonometric functions unique among the elementary functions is their periodicity. This makes them valuable in modelling periodic phenomena which occur in Nature and science.

In general, a periodic function is not sinusoidal (that is, a perfect sine or cosine), but under suitable conditions, it may be approximated by a sum of sinusoidal functions.

4.1 Exercises: Angles and Lengths

1. Neo uses a ramp to load goods onto a vehicle. If the ramp is 5m long, with one end on the horizontal ground and the other 1.5m above ground, what angle does the ramp make with ground? [*See Note 1*] [Ans: $17.46°$]

2. Eona cycles up an incline at a speed of 2 m/s for 2 minutes, gaining an altitude of 10m. Determine the constant angle of the incline. [Ans: $2.39°$]

3. Neo uses a home-made device to measure the angle $45°$ between the ground and the top of a tree when he is 7 m away from the tree. [*See Note 2*]

(a) How tall is the tree? [Ans: 7 m]

(b) How long would the shadow of the tree be when the Sun reaches an elevation of 55°? [Ans: 4.9 m]

(c) Make the simple device Neo used. Use it to measure the height of your school building.

4. Eona determines the height of a tree using a metre rule placed vertically on the ground: She finds that the shadow of the metre rule is 1.3m long when the shadow of her tree is 5.5m long.

   (a) How tall is Eona's tree? [Ans: 4.23 m]

   (b) What was the elevation of the Sun when Eona did her measurement? [Ans: 37.6°]

5. Neo climbs to the top of a lighthouse which is 50 m above sea level. Looking out at sea, he measures the angle of depression to the horizon to be about 0.23°. Use that information to estimate the radius of the Earth. [Ans: $R \approx 6206$ km]

6. Eona wishes to determine the height of building A. She measures the elevation of the roof of A from the base of another building B which is 150 m from A; she finds the angle to be 20°. She then climbs to the top of B and again measures the angle of elevation of the roof of A: Now the angle is 10°.

   (a) What is the height of building A? [Ans: 54.6 m]

   (b) What is the height of building B? [Ans: 28.15 m]

7. A Ferris wheel of radius R has its lowest point a height h above ground. Eona rides in one seat while Neo tracks her position using an angle θ between the vertical diameter and the radial vector to Eona; $\theta = 0$ corresponds to the highest point. Show that Eona's height above ground is given by $H = h + R(1 + \cos\theta)$.

8. Neo and Eona are driving along a straight road. They stop at point A to measure the angle between the road and the direction of a distant mountain to their right: It is 80°. After continuing

to drive for another 20km they stop at point B to again measure the angle between the road and the same mountain: It is now 88°.

(a) What is the horizontal straight line distance from the mountain to the road? [Ans: 141.4 km]

(b) How is the "method of parallax" used to determine the distance of stars from Earth?

9. Eona is on a sailboat, S, some distance from a straight section of the shore. She wishes to estimate how far she is from the shore using angle measurements to two landmarks A and B on the straight section. She measures angle SAP to be 5° while SBP is found to be 10°, P being a point on the shore between A and B which makes SP perpendicular to AB.

   (a) If AB is known to be 500 m long, how far is Eona from the shore? [Ans: 29.2 m]

   (b) Use Eona's method to estimate the diameter of the Moon.

10. A farmer marks his triangular plot of land by planting a tree at each vertex. Tree B is on a bearing 040° from tree A while tree C is located due East of A, at a bearing 135° from B. If $BC = 200$ m, find

    (a) AB. [Ans: 184.6 m]

    (b) AC. [Ans: 260.1 m]

    (c) The area of the triangular plot. [Ans: 18,391 m^2]

11. When light travels from a medium with refractive index n_1 to another with refractive index n_2, such as from air to water, the angle between the light ray and the normal to the interface changes from θ_1 to θ_2. The relationship between the variables is given by Snell's law $n_1 \sin\theta_1 = n_2 \sin\theta_2$. [*See Note 3*]

    (a) Given that the angles are not more than $\pi/2$, show that $\theta_2 > \theta_1$ requires $n_1 > n_2$.

    (b) Find the critical incidence angle $\theta_1 \equiv \theta_c$ which occurs when $\theta_2 = \pi/2$. [Ans: $\sin^{-1}(n_2/n_1)$]

(c) What happens when $\theta_1 > \theta_c$? Discuss one application.

12. The time-varying coordinates of a particle moving in a constant gravitational field (projectile motion) are given by
$$x(t) = Ut\cos\theta \quad \text{and} \quad y(t) = Ut\sin\theta - \frac{gt^2}{2},$$
where U is the initial velocity of the particle, g the acceleration due to gravity, and θ the launch angle relative to the horizontal. (We have ignored air resistance). [*See Note 3 of Chap. 1*]

    (a) Eliminate the time variable to obtain an equation for the trajectory, $y = y(x)$. [Ans: $y = x\tan\theta - (g/2U^2)x^2\sec^2\theta$]

    (b) The range R is the horizontal distance between the start of the projectile and the place where it hits the ground. Find an expression for R. [Ans: $(U^2\sin 2\theta)/g$]

    (c) Determine the maximum value of R as θ varies, keeping the other parameters fixed. What value of θ maximises R? [Ans: $R_{max} = U^2/g$; $\theta = \pi/4$]

    (d) Challenge: Derive the above expressions for $x(t)$ and $y(t)$ starting from $\dfrac{d^2x}{dt^2} = 0$ and $\dfrac{d^2y}{dt^2} = -g$.

    (e) Challenge: How would R_{max} change if air-resistance was not ignored?

4.1.1 Notes

1. Q1: Using a ramp requires less force to raise the object off the ground as only the component of the gravitational force along the ramp (and friction) must be overcome.

2. Q(3, 4): The Greeks used similar methods to measure the height of the pyramids in Egypt. Since the Sun is at a large distance from the Earth, its rays are essentially parallel when they reach us. CAUTION: Never look directly at the Sun.

3. Q11: The speed of light in a medium depends on the properties of the medium (Sect.12.2.2). The refractive index is the ratio $n = c/v$ where c is the speed in vacuum and v the speed in the medium. The frequency of light does not change when

it crosses from one medium to another but its speed and wavelength does. Also, the refractive index depends on the frequency of light, as seen in the dispersion of white light by a prism into its constituent colours, and in the formation of rainbows.

4.2 Exercises: Periodicity and Waves

1. A student observes the angular displacement, θ, of a swinging pendulum. At time $t = 0$ the pendulum is at its maximum displacement of $1/8$ radians from the vertical, and takes 3 seconds to complete a cycle. Assuming the motion is described by the equation $\theta = A\cos(\omega t + B)$, with $0 \leq B \leq \pi$,

   (a) Determine the constants A, ω, B.
   [Ans: $A = 1/8$, $B = 0$, $\omega = 2\pi/3$]

   (b) Re-write the equation for θ in terms of the sine function.
   [Ans: $\theta = \frac{1}{8}\sin(2\pi t/3 + \pi/2)$]

2. Electricity supplied to homes is in the form of "alternating current" (AC). In one country, the formula for the voltage is $V(t) = V_0\cos(2\pi f t)$ where $f = 60$ Hz is the frequency, t the time and $V_0 = 240\sqrt{2}$ Volts. [*See Note 1*]. For the interval $0 \leq t \leq \frac{1}{2f}$,

   (a) Sketch $V(t)$.

   (b) When is the voltage less than 240 Volts?
   [Ans: $1/480 < t < 1/120$]

   (c) When is the voltage greater than -60 Volts?
   [Ans: $0 < t < 0.00464$]

   (d) Why is electricity supplied to homes as AC?

3. A guitar string is tied at its two ends located at $x = 0$ and $x = L$. When plucked, it creates "standing waves" whose vertical displacement is given by $y(x) = A\sin\frac{2\pi x}{\lambda}$ where A is the maximum amplitude and λ is the wavelength of the wave.
   [*See Note 2*]

(a) Using the condition $y(L) = 0$ (since the tied end does not move), find the possible values for λ in terms of a positive integer n. [Ans: $2L/n$]

(b) For $n = 2$, locate the points of the standing wave with zero amplitude (called "nodes") and the points with maximum amplitude (called "anti-nodes"). [Ans: Nodes at $x = 0,\ L/2,\ L$. Anti-nodes at $x = L/4,\ 3L/4$]

(c) What do the words "fundamental" and "harmonics" mean in the above context?

(d) Where else can you find standing waves?

4. An engineer is asked to design a roller coaster track with a sinusoidal shape, $y(x) = A + B\sin(kx)$, where x is the straight line distance as measured along the ground and y is the height of the track above ground. The requirements are: The highest point of the track should be 25m above ground; the distance between the peaks of the track should be 50m; the start of the track at $x = 0$ should be 15m above ground.

   (a) Can you help the engineer determine the constants A, B, k? [Ans: $A = 15,\ B = 10,\ k = 2\pi/50$]

   (b) At what height above ground is the lowest point of the roller coaster track? [Ans: 5 m]

5. When two sound waves of slightly differing frequencies interfere ("add together"), the result is a wave with periodically varying loudness. The perceived variation in the loudness is called a "beat". Representing the two waves by the expressions
   $A_1 = \cos(2\pi f_1 t)$ and $A_2 = \cos(2\pi f_2 t)$,
   with t the time and $f_2 > f_1$ the frequencies,

   (a) Find an expression for $A(t) = A_1 + A_2$ as a product of two sinusoidal functions.

   (b) Plot $A(t)$ for $f_1 = 100$ and $f_2 = 105$ for different ranges of t to discern the shape of the resultant wave. In particular, identify the slowly varying "envelope" of the resultant.

   (c) Show that the resultant wave, $A(t)$, may be interpreted as a sound wave of average frequency $(f_1 + f_2)/2$ whose

amplitude varies periodically.
[Hint: $(f_1 + f_2)$ is of higher frequency than $(f_1 - f_2)$, so the later slower varying component is the "envelope" of the resultant wave $A(t)$.]

(d) If the loudness of sound is proportional to the square of the wave amplitude, explain why the beat frequency is $f_1 - f_2$.

(e) How can the phenomena of beats be used to tune musical instruments?

6. A physicist models the standing waves on a plucked string by the equation

$$A(x) = \sin\left(\frac{2\pi x}{5}\right) - 3\sin\left(\frac{2\pi x}{5} - \frac{\pi}{2}\right),$$

where $x \geq 0$ is the horizontal coordinate and $A(x)$ the amplitude of the wave.

(a) What is the maximum vertical displacement of the waves? [Ans: $\sqrt{10}$]

(b) What is the smallest x at which there is maximum displacement? [Ans: $x = 0.256$]

(c) What is the smallest value of $x > 0$ at which there is no displacement? [Ans: $x = 1.506$]

7. Neo could not sleep well during an overnight camping trip with his fellow Boy Scouts; whenever awake, he would measure the temperature of his surroundings. After his trip he decided to model the day's temperature variation approximately using the function $f(t) = A + B\sin(\omega t)$, with $f(t)$ representing the temperature in degrees Celsius, at time t hours after the start of measurements.

(a) Help Neo determine the constants A, B, ω using the following data he collected: The maximum and minimum temperatures were respectively 32°C and 22°C; the temperature at $t = 3$ was 26°C. (Assume $0 < \omega < \pi/2$).
[Ans: $A = 27$, $B = 5$, $\omega = 1.114$]

(b) According to his model, what was the temperature at $t = 6$? [Ans: 29°C]

8. According to quantum mechanics, which describes the property of matter at atomic scales, any "particle" also has wave characteristics. A particle confined to a one-dimensional box, $0 \le x \le L$, has an associated de Broglie-Schrodinger wave described by $y(x) = A \sin kx$. A is the amplitude, $k = \frac{2\pi}{\lambda}$ and λ is the wavelength. [*See Note 3*]

   (a) Using the condition $y(L) = 0$, find the possible values for λ. [Ans: $\lambda = 2L/n$, n a positive integer]

   (b) The possible discrete energies of the particle in the box are given by $E = \dfrac{\hbar^2 k^2}{2m}$. What is the minimum energy? [Ans: $\dfrac{\hbar^2 \pi^2}{2mL^2}$]

   (c) Are the energy levels of a particle in the box equally spaced? [Ans: No]

4.2.1 Notes

1. Q2: The unit Hertz (Hz) equals one oscillation per second. Over long distances, electrical energy is more efficiently transmitted at very high-voltages, but applications typically require much lower voltages. "AC" has the advantage that its voltage may be stepped-up or down easily using transformers with minimal loss (Sect.12.2.1 and Q8 of Chap.9.5). In this question the "RMS" value of the voltage is 240 V (see Q4 in Chap.7). The frequency and RMS voltages differ from country to country.

2. Q3: In a standing wave the location of the maximum (or minimum) amplitude does not change with time. That is, the wave does not propagate.

3. Q8: $\hbar$ is the Planck's constant h divided by 2π (Sect.13.2.3). See also Q.7 in Chap.6.

Chapter 5

Coordinate Geometry

Points in space or on a plane can be located with the help of coordinates. Examples where this is useful are in navigation and in labelling pixels on a computer screen.

Another important application of coordinate geometry is in the analysis of the relationship between variables. For example, in an experiment one might collect a set of data points (x_i, y_i) for $i = 1, 2, 3....N$, try to deduce a relationship of the form $y = y(x)$ by fitting the data to some theoretical model, and then use the relationship to make some new predictions. In such situations the task is often easier if, through a change of variables from (x, y) to (X, Y), one can get a linear graph: $Y = mX + C$.

5.1 Exercises

1. Eona is touring a city where the streets are laid out in a rectangular grid. She finds herself at a street corner which has map coordinates $(5, 7)$. She wishes to get to a bookshop at location $(10, 19)$. One grid length on the map corresponds to 50 m.

   (a) Estimate how far Eona would need to walk to the bookshop if she followed the streets. [Ans: 850 m]

   (b) If Eona managed to take a shortcut through several buildings and open spaces, what minimum distance would she need to walk to the bookshop? [Ans: 650 m]

2. An investor decides to buy a piece of land for development. He chooses a quadrangle plot by noting the vertex grid coordinates from the seller's map: $(10, 20)$, $(4, 26)$, $(19, 41)$, $(25, 37)$. One unit on the map corresponds to 10 m.

   (a) What is the area of the land chosen by the investor? [Ans: $17,100$ m^2]

   (b) The investor believes he had chosen a rectangular plot of land. Is he correct? [Ans: No]

   (c) If the chosen plot is not rectangular, shift a single vertex to make it so. Specify the new coordinate. [Ans: $(25, 35)$]

   (d) What is the area of the land plot corresponding to part (c)? [Ans: $18,000$ m^2]

3. A toy car moving at the constant velocity 1 m/s is acted at time $t = 0$ by a constant force which causes it to accelerate at 2 m/s^2. The distance travelled by the car, in metres (m), since $t = 0$ is given by

$$x = t + t^2 \ ,$$

   where t is the time in seconds.

   (a) Plot the curve of x versus t.

   (b) Plot the curve of x/t versus t.

   (c) Plot the curve of dx/dt versus t.

   (d) Which of the above curves is a straight line? What is the gradient of the straight line(s)? [Ans: (b) slope 1; (c) slope 2]

4. The number of radioactive nuclei $N(t)$ remaining after time t is given by

$$N(t) = N_0 e^{-\lambda t} \ ,$$

   where N_0 is the number of radioactive nuclei at $t = 0$ and λ is the decay constant. [*See also Q11 of Chap.2*]

   (a) Show that a plot of $y \equiv \ln N(t)$ versus t gives a straight line.

(b) Show that gradient of the line above is given by $-\lambda$ and the y intercept is $\ln N_0$.

(c) Verify that by plotting $\ln\left(\frac{N(t)}{N_0}\right)$ versus t, the new intercept on the y axis is zero.

5. The energy E stored in a stretched spring is given by

$$E = \frac{1}{2}kx^2 ,$$

where k is the spring constant and x is the deviation of the spring length from its unstretched value. Find the gradient and the intercept on the vertical axis if we plot

(a) E versus x^2 [Ans: $1/2k$, 0]

(b) $\ln E$ versus $\ln x$ [Ans: 2, $\ln(1/2k)$]

6. An object with initial speed v is brought to stop in a distance d by a constant opposing force. Theory predicts that $d \propto v^2$. If the opposing force is kept constant while v is varied, suggest a plot involving d and v that would produce a straight line. [*See Sect.11.5*] [Ans: d vs v^2 or $\log d$ vs $\log v$]

7. A manager estimates the cost, in dollars, to produce waffles in his shop using the formula [*See Sect.10.3*]

$$C(x) = \frac{60x + 300}{0.1x + 3} ,$$

where x is the number of batches produced; each batch containing 10 waffles. The sales revenue is estimated to be

$$R(x) = 3x .$$

(a) Re-write $C(x)$ in partial fraction form. [Ans: $600 - 15000/(x + 30)$]

i. What value does $C(x)$ approach as x becomes very large? [Ans: 600]

ii. Sketch $C(x)$.

iii. Does $C(x)$ increase or decrease with x?
[Ans: Increases, reaching the asymptote C=600]

(b) If the profit is simply

$$P(x) = R(x) - C(x) ,$$

determine whether the following values of x's result in a profit or loss: $x = 100$, $x = 200$. [Ans: Loss; profit]

(c) Plot $C(x)$ and $R(x)$ on the same graph. Using the graph, estimate the number of waffles that must be sold to break-even in the business. [Ans: 175.7]

(d) Solve for the last part using analytical methods.

8. A student copied a table showing the distances (a) of the planets from the Sun and their orbital period (T) in some fixed units. Two entries in the table, marked as p and q, were illegible in the student's notes. (a is the semi-major axis of the elliptical orbit). [*See Note 1*]

(a) What units have been used for a and T?
[Ans: They are measured relative to the values for Earth]

(b) Plot $\log T$ against $\log a$ for the data pairs that are complete and estimate the straight line that gives the best-fit. Does it matter which base you use for the logarithm?
[Ans: Any base may be used]

(c) Hence deduce an equation of the form $T = f(a)$, which expresses Kepler's law. Use your equation for the next three parts. [Ans: $T = a^{1.5}$]

(d) Determine the values p, q in the table and compare them with actual values. [Ans: $p = 1.52$, $q = 165$]

(e) If an asteroid is found orbiting the Sun with a period of $T = 2.7$ units, how far would it be from the Sun?
[Ans: $a = 1.94$]

(f) If a spacecraft is in free orbit around the Sun at a distance $a = 25$ units, how long would it take to complete one orbit?
[Ans: $T = 125.2$]

Planetary data

Planet	a	T
Mercury	0.39	0.24
Venus	0.72	0.61
Earth	1.00	1.00
Mars	p	1.88
Jupiter	5.20	11.86
Saturn	9.54	29.46
Uranus	19.19	84.01
Neptune	30.06	q
Pluto	39.48	248.54

9. The intensity of radiation I that penetrates a shielding material of thickness d is given by the relation $I = Ae^{-\mu d}$ where μ is the "attenuation coefficient" and A is the intensity at $d = 0$.

   (a) Show that a plot of $\ln(I/A)$ versus d is expected to be a straight line with slope $-\mu$.

   (b) The data in the table below was obtained using an X-ray source. Which of the two materials, lead or concrete, is the better shield? [Ans: Lead]

   (c) Deduce, using the suggested plot and a line of best fit, the corresponding attenuation coefficients for lead and concrete shielding. [Ans: lead ≈ 2.57 mm^{-1}, concrete $\approx$ 0.0477 mm^{-1}]

   (d) Determine μ using some software and compare with your answer in part (c).

   (e) What thickness of concrete would give the same shielding effect as 2 mm of lead for this X-ray source?
   [Ans: $\approx$ 108 mm]

   (f) A composite shielding is made using 0.5 mm of lead and 50 mm of concrete. With the same X-ray source of intensity A used on one side, what would be the intensity of radiation emerging through the composite shielding?
   [Ans: $I/A \approx 0.025$]

   (g) Where is shielding from X-rays commonly used?

Shielding data

d (mm)	(I/A) (lead)	(I/A) (concrete)
0.5	0.277	0.98
1.0	0.067	0.96
1.5	0.024	0.93
2.0	0.0048	0.91
2.5	0.0017	0.89
3.0	0.00035	0.87
3.5	0.00014	0.85

10. A small mass P connected to the end of a spring undergoes simple harmonic motion along the x-axis. Its position is given by [*See Sect.9.4.3*]

$$x_P = 2\sin 3t \ ,$$

where t is the time in seconds. Another small mass Q is in a similar motion along a line parallel to the x-axis. The position of Q is given by

$$x_Q = 2\sin 6t \ .$$

(a) Plot x_P and x_Q on the same graph for the range $0 \le t \le \pi/3$.

(b) From your plot, find the intersection points of the two curves. [Ans: $t = 0,\ \pi/3,\ \pi/9$]

(c) Find the intersection points using an algebraic approach.

11. A farmer uses 100m length of fencing material to fence up a rectangular area with one side bounded by the river (no fencing on the river side).

(a) Write down an algebraic equation for the bounded area A in this scenario. (Let the side perpendicular to the river be x m long.) [Ans: $A = x(100 - 2x)$]

(b) Plot the area A as a function x and find the value of x which maximises the area. What is the maximum area? [Ans: $x = 25$ m, $A = 1250$ m^2]

(c) Solve the above problem using two other methods: (i) By completing the square, and (ii) using calculus.

(d) More generally, suppose the length of the fence is L, find an expression for the maximised area. [Ans: $L^2/8$]

12. "Laffer's curve" is an attempt to model how a government's revenue $R(x)$, in billions of dollars, varies with the tax rate $x\%$.

    (a) Plot the following possible model curves and find the respective tax rates for which the revenue is maximized.

    i.

    $$R(x) = \frac{1}{200}\left(x^3 - 300x^2 + 20000x\right)$$

    [Ans: 42.265%]

    ii.

    $$R(x) = \frac{400x - 4x^2}{5x + 150}$$

    [Ans: 32.450%]

    (b) Use calculus to re-solve for the above.

5.2 Notes

1. Q8: The planetary orbits are ellipses, with $2a$ the length of the major axis. Kepler's law, with that proportionality constant, applies to any object freely orbiting the Sun. The law may also be applied, for example, to the orbit of the Moon about the Earth but the proportionality constant would then be different. Kepler's empirical law may be derived from Newton's law of gravity (Sect.11.6).

Chapter 6

Differential Calculus

Differential calculus deals with problems involving "rates of change". Such problems are common in science as most of the basic equations describing Nature are differential equations, that is, equations which contain derivatives.

For example, Newton's Second Law of motion, $F = ma$, gives the acceleration, a, of a particle if the forces acting on it are known. If $v(t)$ is the time-varying velocity of the particle, then $a \equiv dv/dt$.

We will sometimes use Newton's over-dot notation for time derivatives, $\dot{x} \equiv dx/dt$ and $\ddot{x} = d^2x/dt^2$.

6.1 Exercises

1. In Q7 of Chap.4.1, if $\theta = \omega t$, where ω is a constant and t the time,

   (a) Find the rate of change of Eona's height H above ground. [Ans: $dH/dt = -\omega R \sin \omega t$]

   (b) At what locations does the height change the fastest? [Ans: $\theta = \pi/2,\ 3\pi/2$]

2. A box of volume 1000 cm^3 is to be constructed as described in Q16 of Chap.1. For what range of values of a is this possible? [Ans: $a \geq 15(4)^{1/3}$]

3. Consider motion along one dimension, as observed from two

frames of reference, one moving relative to the other. As discussed in Q6 of Chap.3, in Galilean-Newtonian physics the coordinates in the two frames are related by

$$t' = t \ ,$$

$$x' = (x - ut) \ ,$$

where u is a constant. Noting that $x = x(t)$,

(a) Evaluate $v' \equiv \frac{dx'}{dt'}$. [Ans: $v' = v - u$ where $v = \frac{dx}{dt}$]

(b) Evaluate $a' \equiv \frac{d^2x'}{dt'^2}$. [Ans: $a' = a$ where $a = \frac{d^2x}{dt^2}$]

(c) Are the results what you expected?
[Ans: $a' = a$ expected]

4. As discussed in Q7 of Chap.3, the Lorentz transformations in one dimensional spacetime are given by $x' = \gamma(x - ut)$ and $t' = \gamma\left(t - \frac{ux}{c^2}\right)$ where $\gamma = 1/\sqrt{1 - u^2/c^2}$.

   (a) Differentiate x' with respect to t.

   (b) Differentiate t' with respect to t.

   (c) Hence show that $v' = \frac{v - u}{1 - \frac{uv}{c^2}}$ where $v' = \frac{dx'}{dt'}$ and $v = \frac{dx}{dt}$

   (d) Challenge: Express $\frac{dv'}{dt'}$, the acceleration in the primed frame, in terms of the acceleration in the unprimed frame.

5. A mathematical biologist models the population of bacteria in a particular culture by [*See Note 4 of Chap.1*]

$$N(x) = -x^3 + 2x^2 + 300x + 4000 \ ,$$

where $0 < x < 20$ is the temperature of the culture in °C.

   (a) Find the temperature at which the number of bacteria reaches its maximum value. [Ans: 10.7°C]

   (b) Show that at that temperature the number of bacteria is indeed a maximum and not just an extremum.
   [Ans: $d^2x/dt^2 < 0$]

6. A ball is thrown upwards with velocity u, starting at an initial height S_0 above ground. Its height after time t is given by

$$S(t) = S_0 + ut - \frac{1}{2}gt^2 ,$$

where g is the acceleration due to gravity. We have defined the upwards direction as positive and ignored air-resistance. All quantities are measured in metric units.

   (a) Find the velocity $v = dS/dt$ of the ball as a function of t. [Ans: $v = u - gt$]

   (b) Find the acceleration $a = dS^2/dt^2$ of the ball as a function of t. [Ans: $a = -g$]

   (c) What is the maximum height reached by the ball? [Ans: $S_0 + (u^2)/(2g)$]

   (d) When does the ball hit the ground? [Ans: $t = (u + \sqrt{u^2 + 2gS_0})/g$]

7. The quantum mechanical wavefunction ψ of a particle confined to a one-dimensional box, $0 \leq x \leq L$ (known as "an infinite square well") is

$$\psi(x) = \sqrt{\frac{2}{L}} \sin\left(\frac{n\pi}{L}x\right) \tag{6.1}$$

where L is the size of the well and $n = 1, 2, 3 \ldots$.

   (a) Evaluate $\frac{d^2\psi}{dx^2}$. [Ans: $-(n\pi/L)^2\psi$]

   (b) Verify that your answer above satisfies the following equation [*See Note 1*],

$$E_n\psi = -\frac{\hbar^2}{2m}\frac{d^2\psi}{dx^2} , \tag{6.2}$$

   where

$$E_n \equiv \frac{n^2\pi^2\hbar^2}{2mL^2} .$$

8. A business manager models the cost of a particular product by

$$C(x) = 10x + \frac{50000}{x+1} + 1000 \ ,$$

where x is the number of items sold (in thousands), while the revenue generated from sales is modelled by

$$R(x) = 100xe^{-(x/100)^2} \ .$$

Use calculus or some software to answer the following:

(a) Find the value of x which minimises the cost.
[Ans: $x = 69.711$]

(b) Find the value of x which maximises the revenue.
[Ans: $x = 50\sqrt{2}$]

(c) The profit is given by $P(x) = R(x) - C(x)$. Find the range of x for which there is no loss. [Ans: $31.509 \leq x \leq 124.481$]

(d) Find the value of x which maximises the profit.
[Ans: $x = 70.571$]

9. A thin-skinned rubber sphere is inflated by pumping it with water (which is essentially incompressible). If the water is pumped in at the rate of 0.02 $\text{m}^3 \ \text{s}^{-1}$, calculate

(a) The rate of change of radius R when the radius is 0.8 m.
[Ans: 0.0025 ms^{-1}]

(b) The rate of change of surface area S when the radius is 0.8 m. [Ans: 0.05 $\text{m}^2 \ \text{s}^{-1}$]

(c) Show that $dS/dt \propto 1/R(t)$.

10. A student wishes to construct a rectangular box of volume V m^3 using cardboard and then decorate its top and four sides (but not the bottom) using sticky plastic sheet. The plastic sheet costs \$C per m^2. If the box must have a square top, determine, in terms of V,

(a) The linear dimensions of the box that would minimise the decoration costs. [Ans: Height $(V/4)^{1/3}$, sides $(2V)^{1/3}$]

(b) The minimum cost to decorate the box.
[Ans: \$3C $(2V)^{2/3}$]

11. A small mass attached to a spring performs simple harmonic motion along the x-axis: Its displacement for $t \geq 0$ is given by $x(t) = 5\cos 2t$. [*See Sect.9.4.3*]

    (a) At what time $t > 0$ does the mass first return to the origin ($x = 0$)? [Ans: $t = \pi/4$]
    (b) What is the velocity of the mass when it first returns to the origin for $t > 0$? [Ans: -10]
    (c) When does the velocity of the mass first become zero for $t > 0$? [Ans: $t = \pi/2$]
    (d) What is the acceleration of the mass when it reaches the situation in part (c)? [Ans: 20]

12. The von Bertalanffy model for tumour growth is given by the equation

$$\frac{dV}{dt} = aV^p - bV^q \,,$$

where V is the size of the tumour, t the time and a, b, p, q are positive constants. [*See Note 2*]

    (a) The two terms on the right hand side of the equation represent growth and degradation. Identify those respective terms. [Ans: aV^p is the growth term.]
    (b) Show that for $q > p$ there is a maximum size to the tumour.
    (c) What is a practical application of such tumour growth models?

13. A biology student finds that the time change of population, $\frac{dx}{dt}$, of bacterial cells in his sample is well described by the logistic equation $\frac{dx}{dt} = Ax(B - x)$, with A, B positive constants. [*See Sect.9.4.2*]

    (a) Determine the stationary points of the population growth. [Ans: $x = 0,\ B$]

(b) What happens when $x > B$?
[Ans: $\dot{x} < 0$, population declines]

(c) Evaluate d^2x/dt^2. [Ans: $A\dot{x}(B - 2x)$]

(d) Sketch the curve $x(t)$.

(e) Suggest an interpretation of the constant B.
[Ans: Stable limiting population]

(f) Discuss another real-world application of the logistic equation.

14. Some animals in a farm become infected with an unknown virus which spreads to the rest. An investigator models the number of infected animals x at time t by the logistic equation, $\dfrac{dx}{dt} = x(a - bx)$, with a, b positive constants.

(a) If the initial rate of change of the infected population, dx/dt, was approximately $3x$, estimate the value of a.
[Ans: $a \approx 3$]

(b) If the epidemic peaked when 1000 animals were infected, estimate the value of b. [Ans: $b \approx 0.003$]

15. If the volume of an inflating sphere changes from 100 cm^3 to 101 cm^3,

(a) Use calculus to estimate the change in its radius.
[Ans: 0.0096 cm]

(b) Use calculus to estimate the change in its surface area.
[Ans: 0.69 cm^2]

(c) Check your results by computing the corresponding quantities for the two spheres exactly.

16. A small object is observed to undergo damped harmonic oscillations along the x-axis: Its displacement for $t \geq 0$ is given by $x(t) = 2e^{-2t}\cos(6t)$. [*See also Q22 below*]

(a) At what time $t > 0$ does the particle first return to the origin ($x = 0$)? [Ans: $\pi/12$]

(b) What is the velocity of the particle when it first returns to the origin for $t > 0$? [Ans: $-12e^{-\pi/6}$]

(c) When does the velocity of the particle first become zero for $t > 0$? [Ans: $t = 0.47$]

(d) When is the acceleration first zero for $t \geq 0$? [Ans: $t = 0.15$]

(e) Sketch the $x(t)$ curve.

17. When a natural oscillator, such as a pendulum, experiences an external driving force which is periodic, its motion can become amplified, leading to the phenomenon of resonance. Consider the driven harmonic oscillator with displacement $x(t)$ satisfying the differential equation $\ddot{x} + a^2x = F \sin bt$, where a, F and b are constants. [*See Note 3*]

    (a) Verify that when $F = 0$, the oscillator equation has the periodic solution $x(t) = A \sin at$ for some constant A.

    (b) For driving force $F \neq 0$, assume a solution of the form $x(t) = B \sin bt$ and determine B explicitly. [Ans: $B = F/(a^2 - b^2)$]

    (c) Show that as the "driving frequency" b approaches the "natural frequency" a, the amplitude B increases.

    (d) In reality, the amplitude cannot increase without bound. What limits its growth? [Ans: Damping]

    (e) What are some applications of resonance?

18. According to the law of radioactivity, the number of nuclei decreases in a geometrical progression with time. The equation which describes this is

$$\frac{1}{N(t)} \frac{dN(t)}{dt} = -\lambda$$

where $N(t)$ is the number of nuclei at time t and λ is the decay constant. [*See Note 4*]

    (a) Suppose the number of nuclei at time $t = t_0$ is N_0, solve for $N(t)$. [Ans: $N(t) = N_0 \exp\left[-\lambda\,(t - t_0)\right]$]

(b) Consider the following decays

$$X \to Y \to Z$$

where the decay constants of X and Y are λ_1 and λ_2 respectively. The differential equation is given by

$$\frac{dN_Y}{dt} + \lambda_2 N_Y = \lambda_1 N_X e^{-\lambda_1 t}$$

where N_Y is the number of nuclei Y at time t and N_X is the initial number of nuclei X. Assuming the number of nuclei Y is zero at $t = 0$, show that

$$N_Y = \frac{\lambda_1 N_X}{\lambda_2 - \lambda_1}\left(e^{-\lambda_1 t} - e^{-\lambda_2 t}\right) .$$

19. Schrodinger's time-independent equation for a particle in a one-dimensional box is

$$E\psi = -\frac{\hbar^2}{2m}\frac{d^2\psi}{dx^2} . \tag{6.3}$$

Solve for the wavefunction $\psi(x)$ with the boundary conditions $\psi = 0$ at $x = 0$ and $x = L$; impose the normalisation condition $\int_0^L |\psi|^2 dx = 1$ to obtain Eq.(6.1) of Q7. [*See Note 1*]

20. The rate of a particular chemical reaction is given by

$$\frac{dx}{dt} = k\,(1 - x)\,(2 - x) \ ,$$

where k is a constant and x is the amount product formed. Solve the differential equation with the initial condition $x = 0$ at $t = 0$. [Ans: $x = \dfrac{2(e^{kt} - 1)}{2e^{kt} - 1}$]

21. An economist models the change in government revenue R (in billions of dollars) with tax rate x (in percentage) by the differential equation

$$\frac{dR}{dx} + \frac{x^2 + 100x - 5000}{(x + 50)^2} = 0 .$$

(a) Determine the revenue R as an explicit function of the tax rate x if $R(0) = 0$. [Ans: $R = \frac{x(100-x)}{x+50}$]

(b) Show that the revenue R is zero at the maximum tax rate $x = 100\%$.

(c) Sketch $R(x)$.

22. Newton's Second Law for a mass under the influence of both a restoring force $F_r = -ky$ and a small damping force proportional to its velocity, $F_d = -c\frac{dy}{dt}$, is given by

$$m\frac{d^2y}{dt^2} = -c\frac{dy}{dt} - ky ,$$

where y is the displacement, m the mass, $c > 0$ the damping constant, and $k > 0$ measures the strength of the restoring force.

(a) Verify that the solution of that differential equation is $y = Ae^{-dt}\cos(\omega t + \phi)$ where A, ϕ are constants that depend on the initial conditions, while d and ω are functions of m, c and k that you should determine.
[Ans: $d = c/2m$, $\omega = \sqrt{4mk - c^2}/2m$]

(b) Sketch the curve $y(t)$.

(c) Check that your solution gives the correct limit when $c = 0$.
[Ans: Simple harmonic motion]

(d) What happens if there is no restoring force, $k = 0$?
[Ans: $y(t) = D + Fe^{-ct/m}$, where D, F are constants]

6.2 Notes

1. Q7: This is the time-independent Schrodinger equation:

$$E\psi = -\frac{\hbar^2}{2m}\frac{d^2\psi}{dx^2} ,$$

where m is the mass of the quantum particle, $\hbar$ is the Planck's constant h divided by 2π and E is the energy of the quantum particle (Sect.13.2.3).

In this exercise, E_n is the energy of the quantum particle confined in a box of width L; n is the quantum number. (See also Q19 above, and Q8 in Chap.4)

The value $(\psi^*\psi)dx$ is the probability of finding the particle in the range dx. "Normalisation" means that the total probability to find the particle within the well is 100%.

2. Q12: This is just one of the many similar approximate models for tumour growth.

3. Q17: In reality, energy loss through some form of resistance will lead to damping of natural oscillations as in Q16. Damping moderates forced oscillations, leading to a finite value of the amplitude at resonance. An application of resonance is in the selection of one radio frequency, from the several available, by tuning the resonant frequency to the desired value.

4. Q18: $\dfrac{1}{N(t)}\dfrac{dN(t)}{dt}$ is the fractional change in the number of radioactive nuclei. It being constant means that the "normalised" rate of decay is a constant (Sect.13.3).

Chapter 7

Integral Calculus

One use of integral calculus is in calculating areas under curves. If the curve $y = p(x)$ represents a probability density, then the area represents the probability for the variable to take values between the integration limits, an application often used in statistics.

More generally, if $y = f(x)$ represents some continuous variable dependent on x, then the average value of y over a range $a \leq x \leq b$ is given by

$$< y > \equiv \frac{\int_a^b f(x)\, dx}{b - a} \tag{7.1}$$

This is a natural generalisation of the mean for discrete variables, $\bar{y} = \frac{1}{N}\sum_{i=1}^{N} y_i$.

Integration may also be viewed as inverting the process of differentiation. Thus it is needed when solving differential equations.

7.1 Exercises

1. Integrate the logistic equation in Q13 of Chap.6 to find $x(t)$ explicitly. Sketch the curve separately for the two cases $x(0) > B$ and $x(0) < B$.

   [Ans: $x(t) = \dfrac{Bx_0 e^{ABt}}{x_0 e^{ABt} - |x_0 - B|}$ where $x_0 = x(0)$.]

2. For the temperature variation in Q7 of Chap.4.2,

(a) Calculate the mean temperature in the time interval $0 \le t \le 6$. [Ans: 27.06°C]

(b) If the mean temperature during the interval $0 \le t \le T$ was 29°C, determine T if $T > 1$. [Ans: 3.63]

3. The number of radioactive nuclei remaining at time t is given by the expression

$$N(t) = N_0 \exp(-\lambda t) \ ,$$

where N_0 is the number at $t = 0$. Find an expression for the mean lifetime τ defined by [*See Note 1*]

$$\tau = \frac{\int_0^\infty t \left|\frac{dN(t)}{dt}\right| dt}{\int_0^\infty \left|\frac{dN(t)}{dt}\right| dt} \ .$$

[Ans: $1/\lambda$]

4. Recall the AC voltage in Q2 of Chap.4.2, $V(t) = V_0 \sin(2\pi f t)$, where $f = 60$ Hz is the frequency, t the time and $V_0 = 240\sqrt{2}$ Volts. The interval $0 \le t \le \frac{1}{2f}$ represents one cycle.

[*See Note 2*]

(a) Determine the mean voltage, $< V >$, over one cycle. [Ans: 0]

(b) Determine the mean of the squared voltage, $< V^2 >$, over one cycle. [Ans: $(240)^2$ V^2]

(c) Determine the "root-mean-squared" (RMS) voltage defined by $\sqrt{< V^2 >}$. [Ans: 240 V]

(d) What is the significance of the RMS voltage?

(e) How is the RMS definition similar to the concept of standard deviation used in statistics?

5. Suppose that the acceleration of an object in a viscous medium is approximated by

$$\frac{dv}{dt} = -kv^2 - v \ ,$$

where v is the instantaneous velocity of the object, t is the time and k is a positive constant. [*See Sect.11.10*]

(a) Find the time taken for the object to reach speed V. You may assume the initial velocity at $t = t_0$ is v_0. [Ans: $t = t_0 + \ln(v_0/V) + \ln((1 + kV)/(1 + kv_0))$]

(b) Express V in terms of t.
[Ans: $V = \dfrac{v_0 \exp(t_0 - t)}{1 + v_0 k - v_0 k \exp(t_0 - t)}$]

(c) Find V when $t \to \infty$. [Ans: 0]

6. The quantum mechanical wavefunction ψ of a particle in an infinite square well is

$$\psi(x) = \sqrt{\frac{2}{L}} \sin\left(\frac{n\pi}{L} x\right) ,$$

where L is the size of the well and $n = 1, 2, 3 \ldots$. [*See Note 3*]

(a) Show that $\int_0^L |\psi|^2 \, dx = 1$.

(b) Find $< x > \equiv \int_0^L x|\psi|^2 \, dx$ when $n = 1$. [Ans $L/2$]

(c) Evaluate $\int_0^{L/2} |\psi|^2 \, dx$. [Ans: 1/2]

(d) The probability to find the particle in a small interval dx located at x is given by $P(x) = |\psi|^2 dx$. What is the probability to find the particle exactly, at say, $x = L/2$? [Ans: 0]

7. A student attempts to model the gas exchange in a certain species of genetically modified plant. She finds that for a short time interval the rate of oxygen production is given by

$$\frac{dP(t)}{dt} = 2t ,$$

where t is measured in hours. Under the same laboratory conditions it's rate of oxygen consumption is

$$\frac{dC(t)}{dt} = 0.5t^2 .$$

If the data is calibrated so that $P(0) = C(0) = 0$, find the time t when the amount of oxygen produced (by photosynthesis) equals the amount of oxygen consumed. [Ans: $t = 6$ hrs]

8. Consider a sphere of radius R. Find the mass of the sphere if its density is given by either of two profiles:

   (a)

$$\rho(r) = \begin{cases} \rho_0 & r \leq R \\ 0 & r > R \end{cases}$$

   [Ans: $(4/3)\pi R^3 \rho_0$]

   (b)

$$\rho(r) = \begin{cases} \frac{\rho_0 r}{R} & r \leq R \\ 0 & r > R \end{cases}$$

   [Ans: $\pi R^3 \rho_0$]

9. An ideal gas is defined by its equation of state

$$pV = nRT \ ,$$

where p is the pressure, V the volume, n the quantity of gas measured in moles, T the temperature, and R is the universal gas constant [*See Sect.11.8*]. The work done by the gas when it expands is given by

$$W = \int_{V_1}^{V_2} p dV \ ,$$

where V_1 and V_2 are the initial and final volumes respectively.

   (a) Find the work done by 1 mole of an ideal gas if it expands at constant temperature T. [Ans: $RT \ln V_2/V_1$]

   (b) A more realistic model for a gas is described by the Van der Waals equation

$$\left(p + \frac{an^2}{V^2}\right)(V - nb) = nRT \ ,$$

where a and b are some empirical constants. Use this model to find the work done at constant temperature by 1 mole of expanding gas. [*See Q14 of Chap.1*]
[Ans: $RT\ln\left(\frac{V_2-b}{V_1-b}\right)+\frac{a}{V_2}-\frac{a}{V_1}$]

10. The centre of mass of a one-dimensional object, such as a rod, is defined by

$$r_{cm}=\frac{\int r\rho\,dr}{\int \rho\,dr},$$

where $\rho(r)$ is the line density of the rod and the integration limits are the ends of the rod. Find the centre of mass of a rod if its line density is given by either of two profiles ($r \geq 0$):

(a)

$$\rho(r)=\begin{cases}\rho_0 & 0\leq r\leq L\\ 0 & r>L\end{cases}$$

[Ans: $L/2$]

(b)

$$\rho(r)=\begin{cases}\frac{\rho_0 r}{L} & 0\leq r\leq L\\ 0 & r>L\end{cases}$$

[Ans: $2L/3$]

11. The acceleration of an object moving vertically near the surface of the Earth is constant, $\ddot{y}=-g$. If its velocity $v(t)=\dot{y}$ and displacement $y(t)$ at $t=0$ are $v(0)=u>0$ and $y(0)=0$, determine [*See Note 4*]

(a) $v(t)$ and sketch the corresponding curve; [Ans: $u-gt$]

(b) $y(t)$ and sketch the corresponding curve. [Ans: $ut-gt^2/2$]

(c) The time when the particle reaches the maximum height. [Ans: $t=u/g$]

(d) The maximum height reached by the particle. [Ans: $u^2/2g$]

(e) The time when the particle returns to $y=0$. [Ans: $2u/g$]

(f) The velocity of the particle when it returns to $y = 0$.
[Ans: $-u$]

12. A thin-skinned rubber sphere is inflated by pumping it with water (which is essentially incompressible). Water is pumped in at the constant rate of 3 cm^3 s^{-1}. If the volume of the sphere is 200 cm^3 when $t = 10$, determine

    (a) The radius of the sphere at $t = 1$. [Ans: 3.457 cm]

    (b) The rate of change of the radius at $t = 1$.
    [Ans: 0.02 cm s^{-1}]

    (c) The rate of change of the surface area at $t = 1$.
    [Ans: 1.7375 cm^2 s^{-1}]

13. Bags are filled with dry cement using a cone. Cement flows into the vertically held right circular cone at its large end and flows out at the narrow end to fill the bags. An applied mathematics student studies the process by creating an approximate model: At time $t > 0$ s, cement flows in at the rate of $10t$ cm^3 s^{-1} and flows out at the rate of t^2 cm^3 s^{-1}. At time $t = 0$ the cone holds 10 cm^3 of cement. [*See also Q20 in Chap.1*]

    (a) Determine the volume of cement, $V(t)$, in the cone at time t. [Ans: $V(t) = 5t^2 - \dfrac{t^3}{3} + 10$]

    (b) Sketch the curve $V(t)$ for $t \geq 0$, locating its local maxima and minima. [Ans: $t = 0$ is a local minimum and $t = 10$ a local maximum]

    (c) At what value of $t > 0$ will the the amount of cement in the cone reach its maximum value V_{max}? What is the value of V_{max}? [Ans: $V_{max} = V(10) = 530/3$ at $t = 10$]

    (d) Determine the value of $t \geq 0$ when the amount of cement in the cone is at a local minimum. [Ans: $t = 0$]

    (e) Beyond which time does the amount of cement in the cone go below the local minimum value found in part (b)?
    [Ans: $t = 15$]

    (f) Will there be a time when there is no cement in the cone? Explain. [Ans: Yes]

(g) If the volume of cement in the cone is given by $V = \frac{h^3}{20}$ cm^3 where h is the height of the cement in the cone, calculate the rate of change of the height of the cement when $t = 5$ s. [Ans: 1.1 cm s^{-1}]

7.2 Notes

1. Q3: The mean lifetime of a radioactive sample can also be interpreted as the time needed for the number of nuclei to decay to $1/e$ of the initial number. See also Q11 of Chap.2.

2. Q4: When a voltage V is applied across a resistor R, the current which flows is V/R (Sect.12.1.3) and this dissipates power $I^2R = V^2/R$ in the form of heat. So the relevant quantity for an electrical heater, for example, is $< V^2 >$ rather than $< V >$ which can be zero.

3. Q6: See also Q7 from Chap.6. The infinite well is a simple model for a confined quantum particle. Here $< x >$ denotes the average value of the position. Since the wavefunction for $n = 1$ (the ground state) is a sine curve, the average value of x will be the middle of the well.

   The integral $\int_0^{L/2} |\psi|^2 dx$ is the probability of finding the particle between $x = 0$ and $x = L/2$. For continuous probability distributions, the probability to find a specific value is always zero.

4. Q11: The acceleration due to gravity is essentially constant near the surface of the earth (Sect.11.6).

Chapter 8

Vectors and Complex Numbers

Information about wind speed is useful but not sufficient in many cases, such as the sprint events at the Olympics: It is also important to know the direction of the wind. Velocity is an example of a vector quantity, which specifies both magnitude and direction.

Vector quantities are commonplace in physics. Examples include acceleration, momentum, force and the electric field. As a result, the rules for adding or multiplying vectors, such as the dot and cross products, were motivated by their usefulness in physical situations.

Complex numbers play a vital role in completing our number system and unifying different parts of mathematics, as seen through the Euler identity (see exercises). They are thus extremely useful in applied mathematics and engineering even though the problems being addressed are in the "real" world.

At a fundamental level, our most successful theory, quantum mechanics, requires complex numbers for its consistent formulation: It is no longer a matter of convenience.

8.1 Exercises

1. Al and Bel are walking with velocities $\vec{v_a} = \left(3\vec{\imath} + 4\vec{\jmath}\right)$ km/h and $\vec{v_b} = \left(3\vec{\imath} - 4\vec{\jmath}\right)$ km/h respectively.

   (a) Find their walking speeds. [Ans: Both 5 km/h]

(b) Find the velocity of Al relative to Bel.
[Ans: $\vec{v} = \left(0\vec{\imath} + 8\vec{\jmath}\right)$ km/h]

(c) What is the angle between their velocity vectors?
[Ans: 106.26°]

2. Given the three vectors

$$\vec{a} = \vec{\imath} + 2\vec{\jmath} + 3\vec{k} \ ,$$
$$\vec{b} = 3\vec{\imath} - 4\vec{\jmath} + 2\vec{k} \ ,$$
$$\vec{c} = 2\vec{\imath} + \vec{\jmath} - \vec{k} \ ,$$

(a) Find the area of the parallelogram with sides $\vec{a}$ and $\vec{b}$.
[Ans: $9\sqrt{5} \approx 20.12$]

(b) Find the volume of the parallelepiped with sides $\vec{a}$, $\vec{b}$ and $\vec{c}$. [Ans: 49]

3. A straight road leads from Al's house to his school located 30 km East and 8 km North relative to his house. Al's father drives him to school, and along the road he picks up Neo who lives in a house 10 km East and 15 km North relative to Al's house.

(a) Find the coordinates of the school relative to Neo's house.
[Ans: 20 km East and 7 km South]

(b) What is the shortest distance between Neo's house and the school? [Ans: 21.19 km]

(c) Neo is picked up at a point on the road which is the shortest distance from his house. What is that shortest distance?
[Ans: 11.92 km]

4. Romeo wishes to row from the South bank of a river to reach Juliet who is on the North bank, 50 m directly opposite. He can row at 1 m/s relative to the river but the river itself is flowing east at 0.25 m/s.

(a) If Romeo attempts to row directly North, what would be his actual (ground) speed, and where would he end up on the North bank. [Ans: Speed of $\sqrt{17}/4$ m/s on a bearing 14.04° East of North]

(b) What heading should he aim for when rowing to reach Juliet exactly? [Ans: 14.48° West of North]

(c) What is the shortest time it would take him to reach Juliet by rowing his boat? [Ans: $40\sqrt{5/3}$ s]

5. In physics, the work W done by a force $\vec{F}$ acting on an object is defined by $W = \vec{F} \cdot \vec{x}$ where $\vec{x}$ is the displacement of the object. Suppose that a force of 5 N is used to pull a cart along the horizontal road. [*See Sect.11.5*]

   (a) If the force is applied at an angle of 30° to the road, what is its component in the direction of motion? [Ans: $5\sqrt{3}/2$ N]

   (b) Write the force in vector form. [Ans: $(5\sqrt{3}\vec{\imath} + 5\vec{\jmath})/2$ N]

   (c) Calculate the work done by the force if the cart is moved 10 m along the road. [Ans: $25\sqrt{3}$ N]

6. Verify that the wavefunction $\psi(x) = e^{-ikx}$ is a solution of the free particle Schrodinger equation [*See Sect.13.2.4*]

$$E\psi = -\frac{\hbar^2}{2m}\frac{d^2\psi}{dx^2}\,, \tag{8.1}$$

   if the energy E is related to the wave-vector k in a certain way. [Ans: $E = \dfrac{\hbar^2 k^2}{2m}$]

7. Verify that the wavefunction of the last exercise also satisfies the equation

$$i\hbar\frac{d}{dx}\psi = p\psi \tag{8.2}$$

   if the momentum p of the particle is related to the wave-vector in a certain way. [Ans: $p = \hbar k$]

8. Euler's identity relates the exponential function to the trigonometric functions, $e^{i\theta} = \cos\theta + i\sin\theta$.

   (a) By multiplying together e^{iA} and e^{iB} obtain the sum rule for the sine and cosine functions.

(b) Use Euler's identity to obtain an expression for $\sin 4x$ in terms of $\sin x$ and $\cos x$. Do the same for $\cos 4x$.
[Ans: $\sin 4x = 4\sin(x)\cos^3(x) - 4\sin^3(x)\cos(x)$,
$\cos 4x = \sin^4(x) + \cos^4(x) - 6\sin^2(x)\cos^2(x)$]

(c) Verify your results in the last part for $x = 0,\ \pi/4,\ \pi/2,$ and π.

Chapter 9

Mathematical Modelling: An Introduction

Many of the exercises in this book illustrate the use of simple mathematical models to describe phenomena. While some of the physical science models are extremely accurate, others are approximations or indicate general trends.

In this chapter we briefly discuss the model building process and its relation to the scientific method.

9.1 The Scientific Method

A scientist seeks to find the most economical and accurate description of phenomena using the scientific method: Testing hypotheses and theoretical models against experiment, or against data from observations of actual phenomena.

The word "model" is usually used to refer to a tentative attempt to explain a certain phenomenon or data. Once a model has been well-tested and developed to sufficient generality, it is often called a "theory".

The scientific method is not a linear process but involves many interlinked and iterative steps in the search for an understanding. Typically, one first has some phenomenon in need of an explanation. The phenomenon might exhibit some regularities which may be summarised by some empirical relations. Experiments (or observations) might be conducted to check the robustness of the phenomena under controlled conditions. A hypothesis may then be formulated: This is just a guess of what causes the phenomena. Predictions can be made based on the hypothesis and further experiments conducted to test them.

Sometimes, more than one model is able to explain the available data without discernible difference. In that case one typically invokes "Occam's razor" to support the simplest explanation over the more convoluted. However, all such support is tentative, until more exploration strengthens our case or causes us to revise our views.

9.1.1 Examples of the Scientific Method

- Since the earliest times we have perceived temporal regularities in Nature: Day to night and day again, the recurring seasons, the motion of the moon and other planets. Ptolemy held that the Earth was the immobile centre of the cosmos with the Sun and other planets revolving around it. Later Copernicus proposed a simpler scheme : The Sun at the centre with the planets revolving in circular orbits around it. The two proposals made different predictions which could be tested once accurate measuring instruments were invented to quantify the observations.

  Later Kepler obtained results, such as the elliptical shape of the orbits, which improved on Copernicus' model but his laws were empirical, without any underlying explanation. That would come only later with the work of Newton and his law of universal gravitation: Kepler's laws are simple consequences of Newton's more fundamental laws which also explain terrestrial phenomena. However even the great Newton had to bow to improvement when Einstein's General Theory of Relativity was verified by experiment and found to give a more accurate account of Nature.

- Neutrinos.
  Around 1930, it was found that the neutron apparently decays according to the formula

$$\text{neutron} \rightarrow \text{proton} + \text{electron} \quad : \ (??) \tag{9.1}$$

  Unfortunately the above process violates energy conservation. One option considered by some physicists of that time was to give up their cherished principle of energy conservation!

However another possibility was suggested by Pauli who postulated the existence of a new particle that he named the neutrino: A third particle on the right-hand-side of the above equation would allow the emitted electrons to have a range of velocities. In order to fit the experimental data, this particle had to be massless, and hence move at the speed of light.

The neutrino was indeed found by Reines and Cowan with the exact properties required to conserve energy.

This example illustrates the scientific method of observing phenomena, forming hypothesis, making deductions, and experimentation. Assuming some principles, such as energy conservation, one can make deductions (an undetected particle carrying away some energy) which can then be tested (experimental searches for the particle).

9.2 Making Estimates

It is often useful to make estimates before constructing a detailed mathematical model.

Making order of magnitude estimates is also useful when we do not have detailed information. For example, how many hairs are on your head? You could count them all, with some effort and much expenditure of time (during which some hairs might fall off). But before starting, could you make an estimate? Is it likely to be in the region of 10^3, or 10^6, or some other number?

For an estimate, you might count the number of hairs in a small region ("sampling"), estimate the area of your head which has hair, and then combine those pieces of information to get your answer. (From the result you would also be able to estimate how long it would take to count all the hairs individually.)

9.2.1 Size and Similarity

Here we illustrate a simple but powerful estimation technique which uses the idea of "scaling" or similarity.

Consider a solid cube of edge length L. Now imagine doubling each edge to $2L$ (keeping the material density the same). Clearly the scaling by a factor of 2 increases the surface area by a factor of 4 and the volume by a factor of 8.

In fact any cross-section of the cube would increase its surface area by a factor of 4, and any three-dimensional region of the cube would increase its volume by a factor of 8, as the edge length doubled.

More generally, by dividing any region into a sum of infinitesimal regions (triangles or cubes), you can convince yourself that when the linear dimensions of any solid are scaled by a factor of k, areas scale by k^2 and volumes by k^3.

If the density remains the same, it would mean that a cube scaled by a factor of k would have k^3 times the weight of the original cube, and the pressure on its base would be $k^3/k^2 = k$ times the original pressure.

Since real materials have finite strength, this means, for example, that scaled up versions of small animals would not be able to support themselves unless their body geometry was changed. This is why we do not see mutant mice the size of elephants, and why elephants have such thick legs.

Here is another example: An ant can easily survive a fall from a great height but not a human. This is because air-drag, which slows down an object, is proportional to the surface area (Sect.11.10) which scales as L^2, but the gravitational force is proportional to the mass of the object and scales as L^3 (for constant density). The larger creature will thus reach the ground with a bigger, and potentially lethal, velocity. (See Q8 of Chap.3).

9.3 Dimensional Analysis

Dimensional analysis is a useful tool for obtaining relationships among the possible variables that might contribute to a measured quantity.

An equation like Newton's second law, $\vec{F} = m\vec{a}$, involves quantities which have different "dimensions": m is a mass, so it is said to have the dimension of Mass; $\vec{a}$ is an acceleration, which can be measured in the units $\mathrm{m/s^2}$ and is said to have the dimension of Length/(Time)^2.

Note that acceleration may be measured in other physical units, such as $\mathrm{km/hr^2}$ but its dimension will still be Length/(Time)^2.

Let us denote the dimensions of Mass, Length and Time by the letters M, L and T, and use the notation $[X]$ to denote the dimension of a quantity X. So, for example, [acceleration] $= LT^{-2}$.

Some quantities are dimensionless, such as angles measured in radians (since that is the ratio of two lengths), and pure numbers such as $-1/2$ and π.

Consistency demands that the dimensions of both sides of an equation match, and this is a quick way of detecting errors. For example, suppose you came across an equation which claimed to give the height reached by a projectile fired upwards with an initial speed v, $h = g/v^2$. Could this equation be correct?

Let us check dimensions. The left hand side has dimensions $[h] = L$ while the right hand side has dimensions $[g/v^2] = [g][v]^{-2} = LT^{-2}(LT^{-1})^{-2} = L^{-1}$. The dimensions do not match, and so the equation cannot be correct.

Notice that the alternative equation $h = v^2/g$ is dimensionally consistent, but this does not imply that it is correct! In fact the correct equation is $h = v^2/(2g)$ (when air resistance is negligible).

So we see that dimensional analysis may be used to detect errors but it does not, by itself, guarantee the correctness of an equation.

Dimensional analysis may be used constructively in the model building process. As an illustration, suppose that we did not know the formula for the period of a simple pendulum and did not know how to derive it. Could we still obtain some information about how the period depends on the possible variables?

We need to first guess the likely variables that might influence the period P: The mass of the bob, m, the length of the string l, the acceleration due to gravity, g, the starting angle from the vertical, A. Let us then suppose a relation of the form

$$P = Cm^w l^x g^y A^z \tag{9.2}$$

where C, w, x, y, z are constants to be constrained by dimensional analysis[1]. Note that $[P] = T$, $[C] = [A] = 1$ (dimensionless). So the right-hand-side of our guessed relation Eq.(9.2) has the dimensions $M^w L^x (LT^{-2})^y = M^w L^{x+y} T^{-2y}$. Since this must equal T we deduce that $w = 0$, $x = -y = 1/2$.

So we conclude that the formula for the period should be of the form $P \propto \sqrt{l/g}$. Even though the proportionality constant might

[1] C might also depend on dimensionless combinations of some of the variables.

depend on the dimensionless amplitude A in some complicated way (in fact it does), we were still able to deduce a simple dependence of the period on the length of the string — without solving a differential equation!

Sometimes one must include dimension-full constants in the analysis, such as the gravitational constant G, whose dimension is determined by Newton's Law of Gravity (see exercises below).

9.4 Creating Mathematical Models

Mathematics is a precise language, with an in-built logic, so it is ideally suited for the scientific study of patterns and trends.

Mathematical models usually start with several approximations and simplifications to capture the essential aspects of the phenomena in a manageable form, and to allow for systematic improvements to the model. (Of course deciding what is essential in a model is an art in itself!)

We illustrate some elementary aspects of the modelling process with three examples. The equations we discuss below appear in some exercises of previous chapters.

9.4.1 Example 1: Finding Correlations

Sometimes we may notice a general trend in a data set and try to quantify our observation using standard techniques (including some guesswork).

For example, Kepler was given a large amount of raw data by Tycho Brahe on the motion of the planets around the Sun. Kepler analysed the data and found some regularities. One clear trend in the data is the increase in orbital period T with the distance R of a planet from the Sun. To quantify this, Kepler probably tried a number of simple relations such as $T \propto R^x$, for x some constant, before reaching the conclusion $T^2 \propto R^3$ (See Q8 in Chap.5).

Finding that relatively simple expression inspired Kepler to search for the underlying basis of his law, the "why"? Unfortunately his attempt was not successful, but some years later Newton achieved that goal.

Sometimes we might have an *a priori* expectation of a particular trend in the data. For example, as discussed above, using dimensional analysis we expect the period of a simple pendulum to scale with its length as $T \propto l^{1/2}$. Then, by an appropriate plot of the data we can deduce the proportionality "constant" and how it varies with the amplitude of oscillations. (Of course in the pendulum case we already have a good quantitative theory which allows us to deduce the relation from first principles).

9.4.2 Example 2: A Logistic Equation

Suppose we wished to describe how the population, x, of rabbits in a locality changes with time t. Since the number of rabbits is an integer, the mathematical expression would take the form of a "difference equation" giving x_n, the number of rabbits at time t_n, for $n = 1, 2, 3, ...$

However, continuous variables are often easier to deal with; so let $x(t)$ represent the number or rabbits per unit area at time t. Treating $x(t)$ as a smooth function, we may use the methods of calculus. The simplest, and seemingly reasonable, assumption to start with would be $dx/dt \propto x$ as this corresponds to $\Delta x \propto x$ during short time intervals: That is, rabbits reproducing at a constant rate.

However, a constant rate of reproduction would lead to an exponential population growth, an unreasonable conclusion since the resources available to the rabbits, such as area and food, are certainly limited. Thus, one expects the rabbit population to face moderating pressures as the population increases.

So let us add a negative term to the right-hand-side of our original proposal, modifying the equation to

$$\frac{dx}{dt} = Ax - Bx^2 , \tag{9.3}$$

with A, B positive constants. When x is small (e.g. 0.1) the linear term dominates and the population increases. However as x increases, the negative term becomes important and moderates the growth. Clearly the growth stops when $Ax - Bx^2 = 0$, and so $x = A/B$ represents a ceiling called the "carrying capacity" of the system. (Note that the carrying capacity in this model is an asymptotic value, reached as $t \to \infty$).

The choice $-Bx^2$ for the moderating term is perhaps the simplest, but it is not the only possibility, nor always the most realistic.

The equation (9.3) is often referred to as the "logistic equation" and the "S" shape of the curve (See Q1 in Chap.7) describes the general trend of many processes, which involve growth and moderation, in the biological and social sciences.

As such processes tend to be quite complicated, with many factors involved, some inspired guess work is typically required for the growth and moderation terms in the equation if one wishes for the model to fit the data reasonably well. Such models are usually called "phenomenological" as they are constructed to fit data rather than derived from first principles.

(See also the Index for questions on the Gompretz function and von Bertalanffy model).

9.4.3 Example 3: Oscillations

Consider a particle moving along the x-axis under the action of a force $F(x)$. Newton's Second Law for this case may be written in the form $F = m\frac{d^2x}{dt^2}$. If the force acting on the particle is a simple restoring force like $F = -kx$, with $k > 0$, then the differential equation describing the motion is $m\frac{d^2x}{dt^2} + kx = 0$.

Setting $\omega^2 = k/m$, you can check that a solution of the equation is given by $x(t) = A\sin(\omega t + c)$ where A, c are constants. This is simple harmonic motion, and is in fact the most general solution of that differential equation with a restoring force[2] $F = -kx$.

Simple harmonic motion is a good approximation to many physical systems which oscillate about an equilibrium point, such as a simple pendulum.

Why is it only an approximation? Well, considering the simple pendulum example for definiteness, the linear restoring force is only an approximation valid for small angles of oscillation. For larger angles the exact expression, which is non-linear, must be considered. We have also ignored air-resistance which damps the motion.

In this example, we could have created a very accurate model from first principles, including the non-linearity and damping, but the

[2]If the force $F(x)$ is a smooth function, it may be Taylor expanded for small displacements, $F(x) \approx a + bx + ...$ If $x = 0$ is the equilibrium point then $a = 0$, and if the equilibrium is stable, $b < 0$ (restoring force).

mathematics would then have become very involved and cumbersome. Although computers can handle very complicated equations, they do not easily yield insights or "rules of thumb".

As mentioned earlier, it is often useful to initially make some simplifications and ignore some factors we believe to be small, so that a manageable model can be constructed. If the simplified model does not fit the data very well, we can always go back and reconsider the simplifications and assumptions we made.

9.5 Investigations

Some questions here are more challenging or open-ended compared to those in previous chapters, while others are actually simpler than they appear. As these questions are meant to encourage exploration and experimentation, we have not provided any answers.

9.5.1 Estimates

1. Make estimates of each of the following:

   (a) The number of hairs on your head.

   (b) The total length of the hairs on your head.

   (c) The number of heartbeats during a human lifetime.

   (d) The number of cell-phones in use in your country.

2. Make estimates of the number of atoms in each of the following:

   (a) Of water in a cup of water.

   (b) In a grain of sugar.

   (c) In the air in your room.

3. Make estimates of each of the following:

   (a) The number of books in your school or public library.

   (b) The amount of space, in Gigabytes, required to store the digital (pdf) version of all the books in part (a).

   (c) The amount of physical space required to store the hard disks containing all those digital books.

(d) Repeat the estimate of the last part, but now applied to the print books existing in the whole world.

4. You might have heard the expression "the human body is mostly water". Use that fact to estimate the masses of each of the following:

   (a) Your heart.

   (b) Your kidney.

   (c) Your liver.

   (d) Compare your estimates with tabulated values.

5. A mountain made of rock of density ρ exerts a pressure $\rho g H$ on its base, where H is the height of the mountain and g is the acceleration due to gravity.

   (a) Estimate the maximum height of mountains on Earth using the limiting value of pressure that the strongest rock can withstand.

   (b) How are mountains formed? What decreases their height?

   (c) Estimate the height of mountains on Mars.

   (d) Compare your estimates with actual heights of mountains on Earth and Mars.

6. How much could your school potentially save in electricity bills if it utilised solar power? How much would it cost to install the solar panels? How long would it take to break even?

7. The exponential function e^x has the infinite series expansion $e^x = 1 + x + \frac{x^2}{2} + \ldots.$

   (a) Use the first few terms of the series to find an approximate solution of the equation $e^x - 2 + x^2 = 0$.

   (b) Use software to solve the equation accurately and compare with your approximate solution.

9.5.2 Dimensional Analysis

1. Neo guesses that the period (T) of a satellite in circular orbit around the Earth depends on the gravitational constant (G), the distance of the satellite from the centre of the Earth (R), and the masses of the Earth (M) and satellite (m).

   (a) Use dimensional analysis to obtain a possible relation giving T in terms of the other variables.

   (b) Compare your expression with Kepler's law.

2. Eona thinks that the escape speed (v) of a rocket from the surface of Earth depends on the gravitational constant (G), the radius (R) of Earth, and the mass of the Earth (M). [Note: Escape speed is the minimum speed required for the rocket to reach "infinity", beyond the pull of the gravitational field]

   (a) Use dimensional analysis to obtain a possible relation giving v in terms of the other variables.

   (b) Look up the actual relation and compare it with your expression.

   (c) Use your relation to explain why it was easier for the Apollo moon missions to leave the surface of our Moon than it was for them to leave Earth.

3. Even when an object moves in a circle at constant speed, its velocity changes because of the change in the direction of motion. Hence the object experiences an acceleration called "centripetal acceleration".

   (a) Use dimensional analysis to deduce how the centripetal acceleration a, depends on the mass of the object m, its speed v, and the radius of the circle r.

   (b) Show that as $r \to \infty$ your expression for a tends to zero. Why is this expected?

   (c) Show that as $v \to 0$ your expression for a tends to zero. Why is this expected?

4. Poiseuille's law gives the rate of smooth flow (volume per unit time) of fluid through a pipe of length L and radius R as

$$f = \frac{\pi P R^4}{8\eta L} ,$$

where P is the pressure difference between the ends of the pipe and η is the "viscosity" of the fluid (which measures its resistance to flow). What are the metric units for viscosity?

9.5.3 Optimisation

(See also the questions in Chap.3 and Chap.6.)

1. You have been asked to investigate two possible ways of packing tomato soup cans in a rectangular box. Each can is a cylinder of height 10cm, with circular cross-section of diameter 7cm. The packing box is to be 10cm high but its width can be adjusted to fit the cans in each case below.

   (a) Case A: 9 cans are to be packed in a box. Draw a picture showing an arrangement which you think would minimise unoccupied space. Determine the volume of your box.

   (b) Case B: 8 cans are to be packed in a box. Draw a picture showing an arrangement which you think would minimise unoccupied space. Determine the volume of your box.

   (c) Calculate the fraction of unoccupied space in each case above.

2. Near a straight stretch of river lie two towns A and B on the same side of the river. A is 50m from the river if one chooses a perpendicular path, while B is 30 from the river along another path perpendicular to the river. Straight line distance between A and B is 300m. Jack starts from A to fill his pail with water at the river and brings it to Jill at B.

   (a) What path did Jack take to minimise his travel distance? (You may assume that there are no physical obstructions).

   (b) Is this problem related to some other problem in physics?

3. The manufacturer of a soya drink is considering three possible packings for his product: A rectangular box, a regular cone, or a regular tetrahedron. The volume of drink to be placed in a packet is fixed.

   (a) Which packing would have the lowest surface area?

   (b) Why does the consideration in part (a) matter?

   (c) What other considerations might affect the choice of packing shape?

4. You have three identical containers, each of volume three litres. The first container is empty. The second contains one litre of a liquid at 90°C while the third has one litre of the same liquid at 50°C. The ambient temperature, T_a is 25°C.
   From Newton's law of cooling, $T(t) = T_a + (T_i - T_a)\, e^{-kt}$, where $k = 0.05$ per minute for this liquid, $T(t)$ is the temperature at time t and T_i is the temperature at time $t = 0$.
   Your objective is to minimise the time it takes for all the liquid to reach 30°C. Explore these possibilities:

   (a) Let the liquid in each container cool by itself.

   (b) Completely mix the liquids so that the starting temperature is the average value 70°C.

   (c) Try some other strategy. (Note: If a volume v_1 of liquid at temperature T_1 is completely mixed with another volume v_2 of the same liquid at temperature T_2, the result is liquid of volume $v_1 + v_2$ at temperature $\frac{v_1 T_1 + v_2 T_2}{v_1 + v_2}$.)

5. A farmer wishes to fence off a rectangular plot of his land, and also fence along one diagonal of the rectangle, to create two adjacent triangular sections for his farm animals. He has L m of fencing and would like to maximise the fenced area. Determine the dimensions of the rectangular plot.

6. A tennis player strikes a ball at an angle θ when it is h m above the horizontal ground. The ball leaves his racket at speed v m/s. If h and v are fixed, determine the equation that must

be solved for θ to maximise the range of the ball. (Ignore air resistance).

7. Neo's sister Eona is standing at the foot of a flat platform which is inclined at an angle α to the ground. She throws a ball up the platform with initial speed u and making an angle β with the platform. What value of β would maximise the range of the ball along the incline? (You may ignore air-resistance).

8. When electrical power (that is, energy per unit time) P is transmitted at a voltage V along a cable of resistance R, the current that flows is given by $I = P/V$. This current results in some energy dissipation in the form of heat. The power loss P_L is given by I^2R.

   (a) Find an expression for P_L in terms of V, P, R.

   (b) If P is fixed, and R has already been minimised, how can power loss be reduced by adjusting V? (Note: The value of V can be adjusted by using transformers).

9.5.4 Modelling

(Other questions on modelling are scattered throughout the earlier chapters.)

1. The Gompertz function $x(t) = ae^{-be^{-ct}}$ is sometimes used to model process which have an initial growth phase followed by a saturation phase.

   (a) Plot the function for $a = b = c = 1$.

   (b) Which section of the curve changes faster: The initial (growth) section or the end (saturation) section?

   (c) Compare your conclusion in the last part with the behaviour of the logistic function in Q13 of Chap.6.

   (d) Suggest one possible phenomenon that might be modelled by the Gompertz function.

2. Use scaling arguments to make estimates of the relative maximum speeds of two fish which have different sizes but the same

shape. Compare your model with others in the literature, and if possible, with real world data.

3. Create a simple physics based model to determine the maximum height a pole-vaulter may reach. Compare with real world data and other models in the literature.

4. Study, through available data, and your own estimates, the distance a car travelling at speed v km/hr would need to stop after the brakes have been applied. Create a model to quantify your analysis. How may your study be relevant for road safety?

5. Study the variation in the height of tides closest to your locality. Can you model the variation approximately, but quantitatively, using a sinusoidal function, or a sum of two sinusoidal functions?

6. Study the daily or seasonal temperature variation in your locality and identify any trends. Attempt to create simple mathematical models that quantify the trends. Compare your efforts with other analyses.

7. Identify one acute problem facing your school or city. Gather data to quantify the problem. Analyse probable causes and potential solutions. Attempt to create a model which would simulate how your proposed solution would alleviate the problem. Present your proposal to a critical audience and improve your model, if necessary or possible, based on the feedback received. Obtain approval to test your proposal. Good luck!

9.5.5 Other Questions

1. A body at temperature T Kelvins (K) loses heat by radiation. If the surroundings are at a constant temperature T_o K, the net power loss per unit surface area is given by

$$H = e\sigma(T^4 - T_o^4) \tag{9.4}$$

where $0 < e < 1$ is the emissivity of the body and $\sigma = 5.67 \times 10^{-8}$ W m^{-2} K^{-4} is the Stefan-Boltzmann constant. The Kelvin scale temperature T equals $\theta+273.15$ where θ is the temperature in degree Celsius.

(a) Calculate the power loss by radiation for a sphere of radius 0.5m at temperature 50°C. Take $e = 0.5$ and the ambient temperature to be 25°C.

(b) A cube, made from the same material as the sphere, is found to have the same power loss at the same ambient temperature. If the edge length of the cube is the same as the diameter of the sphere, determine the initial temperature of the cube.

2. Eona runs a distance d at speed v_1 and covers the same distance on her return trip at speed v_2.

   (a) What was her average speed $< v >$ over the distance $2d$? ($< v >$ is the total distance travelled divided by the total time taken.)

   (b) Prove that $< v >$ must be smaller than both $2v_1$ and $2v_2$.

3. In the note to Q17 in Chap.6 it was stated that damping would lead to a finite value for the forced oscillations. Verify that claim by solving the differential equation for forced harmonic oscillations, including a damping term proportional to the velocity.

Chapter 10

Money in a Nutshell

10.1 Interest

Money deposited in a bank savings account typically earns interest. "Simple interest" refers to a fixed rate that is paid on the principal amount, so that simple interest at $R\%$ per annum on a principal of P_0 dollars will create $P_0(1+r)$ dollars at the end of the year, where we have denoted $R/100 = r$.

Interest is often compounded: The nominal annual interest of $R\%$ may be divided into N parts, and an interest of $R/N\%$ paid on the accumulated amount (principal plus interest) at the end of each $\frac{12}{N}$ months. So at the end of t years an initial savings of P_0 would have grown to

$$P(t) = P_0 \left(1 + \frac{r}{N}\right)^{Nt} .$$

Theoretically, if the compounding occurs continuously, then one may take the limit $N \to \infty$, and since

$$\lim_{N\to\infty} \left(1 + \frac{1}{N}\right)^{N} \to e ,$$

so

$$P(t) \to P_0 e^{rt} .$$

We see that compounding several times a year makes the savings grow faster than a simple annual interest. The effect of compounding may be summarised by quoting an "effective annual interest rate", which is the simple annual interest that gives the same result as compounding.

10.2 Supply and Demand

The quantity of a good or service demanded by consumers is affected by its price and a number of other factors, such as the availability of alternatives. The "law of demand" is a hypothesis which says that, all other factors being the same, the quantity q_d in demand is negatively correlated with its price p. A linear model is

$$q_d = a - bp \ ,$$

with $a, b > 0$.

Likewise, the "law of supply" is a hypothesis which says that amount q_s of the product that is in supply is positively correlated with the price, when all other factors are the same. In the linear model,

$$q_s = cp - d \ ,$$

with $c, d > 0$.

Market equilibrium refers to the state when the supply of goods meets demand, $q_s = q_d$, or equivalently $p_s = p_d$.

Although the linear models are convenient to use, it is important to remember that they are simple approximations. Certainly any smooth curve may be approximated in a small interval by a straight line, but extrapolating along the straight line might not be very accurate.

Also, note the "all other factors being the same" assumption which was used to simplify the discussion.

10.3 Cost and Revenue

The cost function involves two components which express the manufacturing costs of a product. Firstly, there are "fixed costs" which involve factors such as product design, the setting up of a factory and training workers.

The second component of the cost function expresses "variable costs", such as the quantity produced, the cost of raw materials, and shipping charges.

A simple model for the cost function is therefore

$$C(q) = c_1 + c_2 q \ ,$$

where c_1 is the first contribution mentioned above and q the quantity of goods produced.

The revenue generated, $R(q)$, is determined by the price p per item. If the price is constant,

$$R(q) = pq \ .$$

A simple profit function is

$$P(q) = R(q) - C(q) \ ,$$

and the break-even point occurs when $P(q) = 0$.

Bear in mind that the model above is highly simplified and is only meant to illustrate certain approximate trends.

By combining what we have discussed so far, we can display an explicit non-linear model for the revenue $R(q)$. Since $R \propto pq$ where q is the quantity sold, and since the quantity in demand may be modelled by $q \approx a - bp$, combining those two equations leads to $R \propto p(a - bp)$, a parabola with a maximum; the revenue rises at first but then decreases after a certain point. The revenue is zero not only at $p = 0$, as expected, but also at some $p^* = a/b$.

Chapter 11

Classical Physics in a Nutshell

11.1 Measurement and Units

The metric units for length, mass and time are the metre (m), kilogramme (kg) and second (s). The prefix "kilo" stands for 10^3; one kilogramme is 1000 grammes and one kilometre is 1000 metres. Other common prefixes for large quantities are "Mega" (10^6) and "Giga" (10^9).

For small quantities the common prefixes are "milli" (10^{-3}), "micro" (10^{-6}) and "nano" (10^{-9}).

Some derived units of measurement have their own unique abbreviation. For example, since from Newton's Second Law force is mass times acceleration, the unit for force is kg m s^{-2}, which is called the "Newton" (N).

Avogadro's constant, N_A, is 6.022×10^{23}; it is the number of atoms in 12 grammes of Carbon-12. The "mole" is a unit used in chemistry to express the amount of substance which has N_A constituents. So, 1 mole of Carbon-12 has a mass of 12g, while one mole of water has N_A molecules with a total mass of 18g.

The standard (or scientific) notation is very useful for expressing numbers compactly. Any non-zero number may be written in the form $\pm A \times 10^n$ where $1 \leq A < 10$ and n is an integer (positive or negative).

The number of digits in the "A" of the standard notation expresses the "significant digits" for that number. For example, the distance from the Earth to the Moon is about 384,000 km. This may be written as 3.84×10^5 km or 3.84×10^8 m, expressing the distance correct to 3 significant digits (if we are confident of the value). If we were less certain of the last digit, we could express the distance as 3.8×10^8 m, signifying a precision of 2 significant digits.

The power, n, of 10 in the standard notation is useful for comparing the relative sizes of numbers. If two numbers of the same sign have the same power in standard notation then we say they are of the same "order of magnitude". One order of magnitude is one power of ten; a number with $n = 5$ is two orders of magnitude larger than a number with $n = 3$.

For example, the distance of the Sun from Earth is about 1.5×10^8 km, which is about 3 orders of magnitude larger than the distance of the Moon from Earth.

11.2 Kinematics

Kinematics is the study of motion, without enquiring about the causes of motion. To describe the motion of an object through space we need to quantify both the location of the object and the passage of time.

In Newtonian mechanics, time is an absolute (observer independent) quantity that can be measured by some reliable clock. Space is taken to be Euclidean, with fixed properties, and is in this sense absolute.

An object is referenced with respect to some frame, which has a coordinate system associated with it, so that at any time we may label the position of the object (taken for convenience to be point like). The simplest coordinate system is then the Cartesian one, the particle's position being given by $x(t), y(t), z(t)$ at any moment in time. One may then calculate the particle's velocity and acceleration from the knowledge of its trajectory.

It is important to note that coordinate systems do not have an independent physical existence; they are constructs we use to describe motion and we are free to choose any convenient coordinate system. The physical phenomenon cannot depend on the coordinate system chosen by a particular observer, though the mathematical description may appear different for different choices (for example, describing the flight of an aeroplane in Cartesian or spherical coordinates).

Speed is defined as the rate of change of distance travelled, ℓ, with time, $s = \frac{\Delta \ell}{\Delta t}$, where we use the symbol Δ ("delta") to represent a change; so Δx means change in x. If the object travels equal distances in equal times, then its speed is a constant and equals its average speed. However, in general the speed of an object changes during its motion and so we must consider its *instantaneous* speed at any particular time. This is obtained by calculating s in a small time interval Δt, and then taking the limit $\Delta t \to 0$. In that limit we use the calculus notation, $s = \frac{d\ell}{dt}$.

Velocity defines not only speed but also the direction. Since direction can be specified by a vector $\vec{x}$, velocity is also a vector quantity, given by $\vec{v} = \frac{d\vec{x}}{dt}$. So constant velocity means not only constant speed but also constant direction. A ball tied to the end of a string can be made to travel at constant speed around a circle but since its direction of motion is changing, its velocity is not constant.

The components of velocity along the three Cartesian axes are denoted as v_x, v_y and v_z. The SI units for both speed and velocity are m/s.

All frames of reference moving at constant velocity relative to one other are indistinguishable: One cannot perform experiments within such frames to discover one's motion (velocity).

Consider two frames of reference, as shown in Fig.(1.1) of Chap.1. The first, referred to as the S frame, is what we will typically think of as the "rest frame". It will have the spatial coordinate system (x, y, z) with origin O and an observer called Stan. The second frame, S', with observer Maria, has coordinate system (x', y', z') and moves with constant velocity u relative to the S frame in the x-direction, with its y' and z' axes parallel to the y and z axes respectively. We further assume that at time $t = 0$ the two origins O and O' coincide.

Since time is absolute in Newtonian physics, so $t = t'$. Consider a point P with coordinates (x, y, z) in the S frame. Then its coordinates in the S' frame will be $y' = y$ and $z' = z$ but since S' moves to the right we have $x' = x - ut$. The set of equations

$$t' = t \tag{11.1}$$

$$x' = x - ut \tag{11.2}$$

$$y' = y \tag{11.3}$$
$$z' = z. \tag{11.4}$$

are referred to as the Galilean transformations between the coordinates of the two frames. From these one may deduce the velocity and acceleration transformation rules. For example, if a particle moves from point 1 to point 2 then from (11.2) we have $x_1' = x_1 - ut_1$ and $x_2' = x_2 - ut_2$. Subtracting those equations gives $\Delta x' = \Delta x - u\Delta t$ and[1] then dividing by Δt gives

$$v_x' = v_x - u \tag{11.5}$$

and similarly $v_y' = v_y$, $v_z' = v_z$.

Acceleration is the rate of change of velocity with time: $\vec{a} = \frac{d\vec{v}}{dt}$. So acceleration is also a vector, with units m s^{-2}.

11.3 Newton's Laws

Newton formulated three laws of motion:

1. *First Law: In an inertial frame, an object remains at rest, or continues its motion at constant velocity, unless acted on by an external force.*

   The first law helps you identify some special frames of reference called "inertial frames": If you cannot find any interaction between a body and its surroundings (that is, no forces act on the body) and yet you observe the body to change its velocity, then you are not in an inertial frame.

   In other words, in an inertial frame, an object changes its velocity only when acted on by some force.

2. *Second Law: The force acting on an object is equal to the product of its mass and acceleration,* $\vec{F} = m\vec{a}$.

   Newton's Second Law tells us how much force is needed to cause a certain acceleration, or conversely what acceleration would

[1] We denote $x_2 - x_1 = \Delta x$ etc.

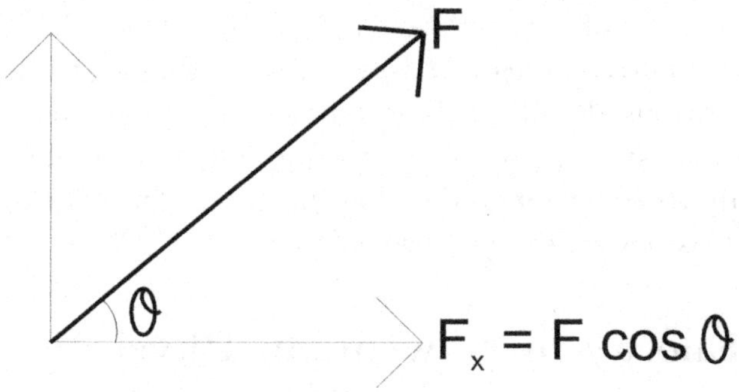

Figure 11.1: Resolution of Forces (finding component along an axis).

result if a given force acted on an object. The parameter m appearing in the equation is the *inertial* mass of the object, a measure of the object's inertia or resistance to change. We will simply refer to the inertial mass as mass.

The SI unit of mass is the kilogramme, kg, and so the unit of force is kg m s^{-2}. This basic unit is given the name Newton.

3. *Third law: If one object exerts a force on a second object, then the second object exerts an equal but opposite force on the first object.*

11.4 Conservation Laws

Advanced studies of kinematics show that continuous symmetries of a physical system give rise to conservation laws.

"Symmetry" means that something (geometrical or mathematical) does not change its form when we carry out an operation such as moving the system to the right or rotating it.

If the laws of physics for a system do not change with time, then associated with the system is a conserved quantity called "energy". The total energy in a closed system does not change though it may transform into another form. For example, the potential energy of a falling rock changes to kinetic energy and then to sound and heat on impact with the ground. The Law of Energy Conservation is one of the most important laws of nature.

The existence of conservation laws allows us to place constraints on the possible dynamics of a system even though we might be ignorant of various details; this is particularly important in the case of complicated systems with many components whereby it might be too difficult to solve for and follow in detail the trajectory of each component as described by Newton's laws.

11.5 Energy in Newtonian Physics

In physics, the *work*, W, done by a force F on an object is defined by

$$W = \vec{F} \cdot \vec{d} \tag{11.6}$$

where $\vec{d}$ is the displacement of the object. The unit of work is therefore the "Newton-metre", which is also called the Joule.

For example, a stone released from the top of a mountain will roll down the mountain gaining speed, and hence kinetic energy, T (energy due to motion): The gravitational force did work on the stone, and this work increased its kinetic energy.

The formula for the kinetic energy of an object of mass m moving at (*non-relativistic*) speed v is given by

$$T = \frac{1}{2}mv^2. \tag{11.7}$$

Energy due to position or the internal configuration of a material body is called potential energy, for example the energy stored in a compressed spring is called elastic potential energy. A stone on a mountain top has gravitational potential energy. A charged particle stationary in an electric field has electrostatic potential energy.

The time rate of transfer of energy is called Power. Its metric unit (Joules per second) is given the name "Watt", W.

Newton's Second Law can be re-written in the following form:

$$\vec{F} = \frac{d}{dt}(m\vec{v}) \tag{11.8}$$

with $\vec{v}$ the velocity of the particle. The quantity $p = m\vec{v}$ is called the (linear) momentum of the particle, and so Newton's law states that the force acting on a particle is equal to the rate of change of its momentum.

11.6 Newton's Law of Gravitation

Newton's Law of Universal Gravitation states that the force of attraction between two point masses (m, M) is directly proportional to their masses and inversely proportional to the square of the distance between them:

$$\vec{F} = -\frac{GMm}{r^2}\,\hat{r}\,, \tag{11.9}$$

where $G = 6.67 \times 10^{-11}$ N m^2/kg^2 is the gravitational constant and $\hat{r}$ a unit vector in the radial direction pointing from the centre of the mass M. The law (11.9) also applies to spherical masses, with r the distance between the centres of the masses.

If an object is released above the Earth, Newton's Second Law together with the law of gravity (11.9) imply that the acceleration of the object is $a \equiv g = -\frac{GM}{r^2}$, independent of the particle's mass, as Galileo found, but depending on r. Near the surface of the earth one can put $r \approx R_E$, the radius of the earth, and get the usual value for g of about 9.8 m/s^2.

11.7 Buoyancy and Archimedes' Principle

Unlike solids, fluids such as air and water do not have a fixed shape but adapt to the shape of the container or surroundings they are in.

The upward buoyancy force acting on an immersed (partially or fully) object is equal to the weight of the fluid displaced. This is known as Archimedes' Principle.

If ρ_0 is the density of the fluid, V_0 the displaced volume and g the acceleration due to gravity, the upthrust is equal to $\rho_0 V_0 g$. On the other hand, the downward force on the object is equal to its weight, which is $W = \rho V g$, where V is the volume of the object and ρ its density. An object will float only if the upthrust can balance the weight; this can only happen if $\rho = \rho_0 V_0/V \leq \rho_0$.

11.8 Gases

Real gases at moderate pressures and temperatures obey a universal gas law which can be summarised by the equation $PV = a(\theta + \theta_0)$, where P is the pressure of the gas (force per unit area), V its volume, and θ is its temperature on the Celsius scale where 0°C is the freezing point of water. a and θ_0 are universal constants.

An extrapolation of the empirical curve to the limit $PV = 0$, gives the intercept $\theta_0 = 273.15$. It is convenient to define an absolute temperature scale T, measured in Kelvins, K, by

$$T = \theta + 273.15\,. \tag{11.10}$$

The ideal gas law is then defined by the equation

$$PV = NkT\,, \tag{11.11}$$

where $k = 1.38 \times 10^{-23}$ J/K is the Boltzmann constant and N the number of atoms/molecules in the gas.

The lowest temperature is 0 K, absolute zero, where all molecular motion would cease; however quantum fluctuations prohibit one from reaching absolute zero: This is codified as the Third Law of Thermodynamics.

11.9 Water Pressure

Unlike air, water is an incompressible fluid. This means that pressure applied to water at one part of an enclosed system gets transmitted to other parts. The incompressible nature of water also makes the calculations of the effects of water pressure easier. For example, the pressure due to a column of water of height h is clearly given by

$$P = \rho g h \tag{11.12}$$

where ρ is the density of water and g the acceleration due to gravity. Since the density of water is 1000 kg/m^3 and $g = 9.8$ m/s^2, the pressure due to a 10m water column is about $100,000$ Pa, which is normal atmospheric pressure (1 Pa = 1 N/m^2). Thus, in water the pressure increases by about $10,000$ Pa for every 1m of depth. Hence, 10m below sea-level the pressure on an object is about 2 atmospheres.

11.10 Drag

There are various forms of drag, or resistance to motion, that an object can experience in a fluid. In general it is neither possible nor convenient to consider the separate forms of drag, so one usually just defines a drag coefficient for a particular shape. The drag F_D (force opposing motion) is related to the drag coefficient, C_D, by the relation

$$F_D = C_D\, A\, \frac{\rho v^2}{2}\,, \tag{11.13}$$

where A is the projected area of the body on a plane perpendicular to the direction of fluid flow, ρ is the density of fluid and v is the flow velocity. The drag coefficient depends on the shape considered and also on the speed (or Reynolds number).

For small objects moving at very low speeds, one may use the approximate relation $F_D \approx v$ instead of the previous equation. Sometimes the alternate relation $F_D \propto -v - kv^2$ is used for intermediate ranges of speed.

Due to the complexity of fluid dynamics, most of the expressions for the drag force represent simplified models or empirical approximations.

Chapter 12

Electromagnetism in a Nutshell

12.1 Electricity

12.1.1 Charges

Matter at the microscopic level consists of building blocks called atoms. A crude model of the atom is that of a number of electrons orbiting a core called the nucleus. The electrons are held in their orbit by the electrostatic force of attraction between unlike charges: The electrons are negatively charged while the nucleus is positively charged.

Coulomb's law states that the magnitude of the force between two point charges Q_1 and Q_2 separated by distance R in vacuum is given by

$$F = \frac{k_c Q_1 Q_2}{R^2} \tag{12.1}$$

where $k_c = 9 \times 10^9$ N m^2/C^2 is a universal constant, and the direction of the force is along the line joining the two charges.

The unit for charge is Coulomb (C). Electric charge comes in multiples of the electron charge, 1.6×10^{-19}C. Charge is conserved: That is, the total charge in a closed system is fixed.

12.1.2 Fields

Any electric charge generates an electric field which then acts on other charges. The field of a positive charge creates an attractive force on negative charges and a repulsive force on positive charges. The electric field is described by a vector assigned at every point in space.

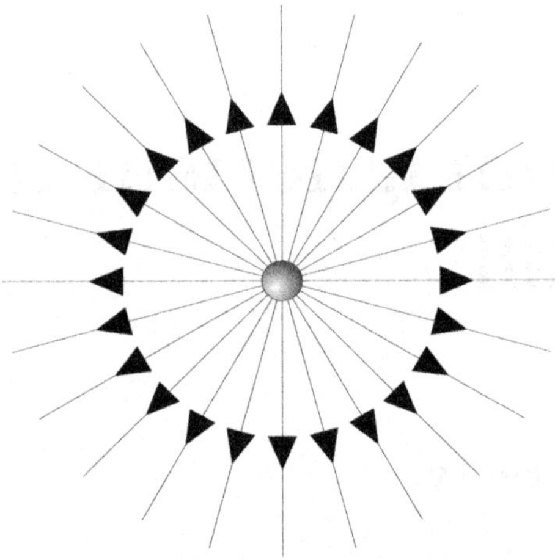

Figure 12.1: Electric field due to a positive point charge

Mathematically, the electric field produced by a point charge is defined as the force it produces on a unit positive test charge. Therefore, an electric field $\vec{E}$ exerts a force $\vec{F}$ on a charge Q given by

$$\vec{F} = Q\vec{E}\ . \tag{12.2}$$

Thus the SI unit for electric field is Newton/Coulomb (N/C).

12.1.3 Potential

In an earlier chapter, we encountered the concept of potential energy, an example being gravitational potential energy — it requires work, done against the force of gravity, to lift an object a certain height, and the object is then said to have gained potential energy.

Similarly, instead of talking of electric forces and fields which are vector quantities, one can introduce the concept of electric potential difference between two points, defined as the work required to move a unit (one Coulomb) positive test charge between those two points.

Thus the unit of electric potential is Joule/Coulomb and this is simply called the Volt (V).

A battery is a device that creates potential differences through chemical means. Chemical potential energy in the battery is used to drive electrons from the positive terminal to the negative terminal and this continues until the electrostatic potential difference becomes large enough for equilibrium to be reached. At this point, small dry cells have a potential difference of 1.5 V between their two ends.

If a battery is connected by a piece of wire, the potential difference between the two ends causes a flow of electrons: The chemical energy stored in the battery is converted to the energy of the electrons. If one Coulomb of charge flows past any point in one second, then one says that a current of one Ampere (A) is flowing.

Real wires resist the electrical current to varying degrees, due to the collision of the electrons with the atoms of the wire. If I is the current which flows through a wire when the ends are at a potential difference of V, then Ohm's law gives the resistance as $R = V/I$.

Electrical resistance causes dissipation of energy as heat.

12.2 Magnetism

In addition to electric charges and forces, there exist analogous magnetic forces.

All magnetic materials always have two opposite poles: A north pole and a south pole, with like poles of materials repelling and unlike poles attracting. Just as we introduced the electric field, we also have the concept of magnetic field, denoted by $\vec{B}$. The magnetic field lines around a magnet are shown in Fig.(12.2).

12.2.1 Electromagnetism

Electric currents produce magnetic fields. The creation of magnetic fields by currents is behind the operation of electromagnets which are found in numerous modern appliances such as the electric motor.

Conversely, a changing magnetic field produces electric fields. This process is called electromagnetic induction. If the magnetic flux

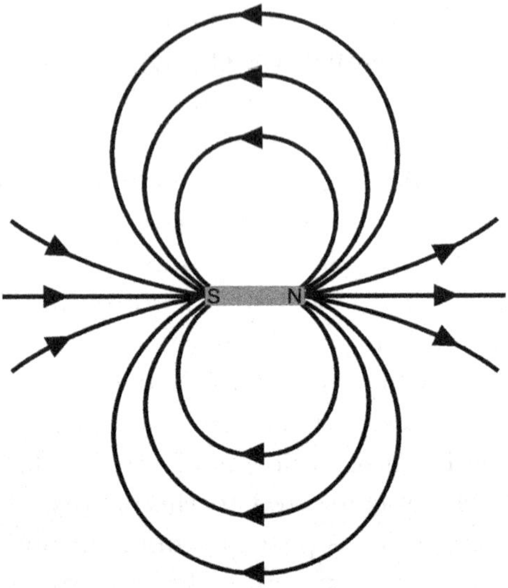

Figure 12.2: Magnetic field lines due to a magnet

through a coil is changed, the induced electric fields produce currents which tend to oppose the change (Lenz's law).

Electromagnetic induction is responsible for the functioning of electrical generators which supply us with electric power. A simple generator has a rotating coil in the field of a permanent magnet. As the coil rotates, the flux through the coils changes, inducing an electric current. However the direction of the induced current changes during each half cycle to satisfy Lenz's law. This kind of current is called an alternating current (AC), in contrast to the unidirectional "direct current" (DC) provided by a dry cell.

12.2.2 Electromagnetic Waves

The fundamental laws of electricity and magnetism as summarised by Maxwell's equations describe the unified phenomenon of electromagnetism. Maxwell's equations satisfy the principles of special relativity.

Maxwell deduced that oscillations of the electromagnetic field gives rise to electromagnetic (EM) waves. In contrast to sound waves, for EM waves there is no underlying material medium whose oscillations result in the wave phenomenon: That is, EM waves can propagate in a vacuum.

Another important difference between EM waves and sound is that EM waves are transverse waves, in contrast to sound which is a longitudinal wave. That is, the wave oscillations for EM waves are perpendicular to the direction of propagation of the wave, whereas for sound waves the oscillations are in the same direction as the propagation of the wave.

All electromagnetic waves travel at the same speed through vacuum: At the speed of light, which is about 3×10^8 m/s.

The wavelength of a wave is the distance between two points on the wave which are in phase, for example, the two peaks in the propagation of the wave. Frequency is the number of times the wave oscillates each second. So for a propagating wave, the speed (c) is related to the wavelength (λ) and frequency (f) by the relation

$$c = \lambda f \, . \tag{12.3}$$

Light is just a special case of electromagnetic waves: Those with wavelengths in the range of 10^{-6}m to 10^{-7}m form what we call the visible spectrum.

When waves propagate from one medium to another, their frequency remains unchanged while their wavelength changes, so that the speed of EM waves in a medium actually depends on their frequency. For example, white light splits into several colours when passing through glass because the low frequency components (red) travel faster in glass than the high frequency (blue) ones.

(OPTIONAL: In one dimension, x, the equation describing wave propagation is of the form

$$\frac{\partial^2 f(x,t)}{\partial x^2} - \frac{1}{v^2}\frac{\partial^2 f(x,t)}{\partial t^2} = 0 \, , \tag{12.4}$$

where $f(x,t)$ is the changing quantity characterising the wave (e.g. pressure amplitude for sound or electric field for EM waves) and t is the time variable. Here $\frac{\partial}{\partial x}$ is the "partial derivative" with respect to x, keeping t constant and likewise for the other case and higher order derivatives.)

It is easy to check by substitution that

$$f(x,t) = A \sin 2\pi\left(\frac{x}{\lambda} - \frac{t}{T}\right) , \tag{12.5}$$

is a solution of the wave equation if $v = \lambda/T \equiv \lambda f$ (the constant A is the amplitude of the wave). From the form of the solution one sees that λ is the wavelength of the wave, that is, the distance between two points on the wave that are in phase, while T is the period or time taken for the wave at a fixed location to repeat its value. The frequency $f = 1/T$ is therefore the number of oscillations of the wave in one second and

$$v = f\lambda, \tag{12.6}$$

is the wave speed.

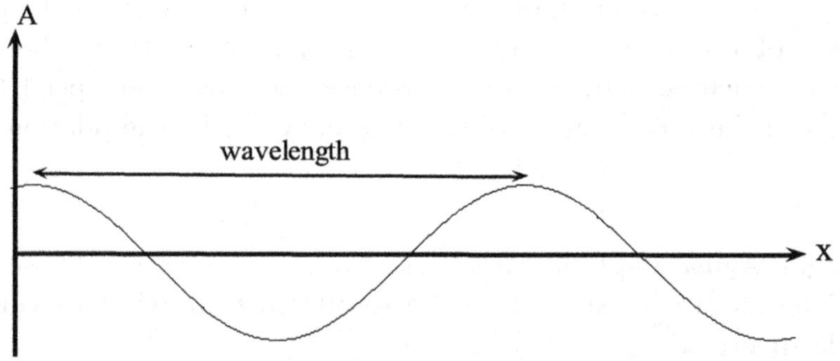

Figure 12.3: Definition of Wavelength

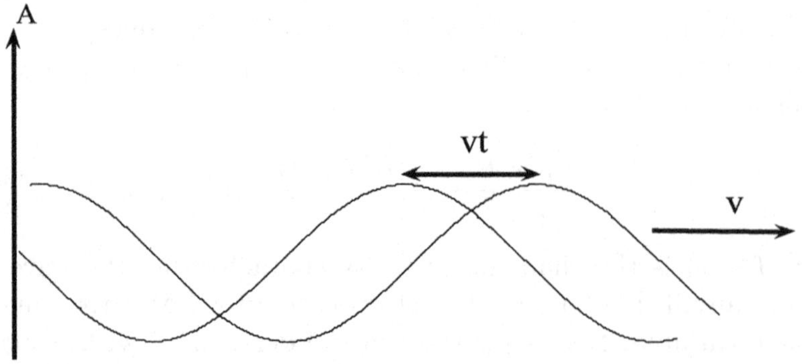

Figure 12.4: Wave moving at speed v

Chapter 13

Modern Physics in a Nutshell

13.1 Special Relativity

Although Newtonian physics was immensely successful in describing the mechanical world, there were other branches of physics which required new concepts and ideas. As we saw, Maxwell's equations predicted the existence of electromagnetic waves that travelled at a speed c in a vacuum. Such waves were detected, and it was confirmed that light was an example of an electromagnetic wave.

According to Galilean transformations, Eqs.(11.1), the speed of light was expected to be different relative to moving observers; $c \pm u$ relative to a frame moving parallel to the direction of light. However, all experiments showed that the speed of light in vacuum was an invariant, always equal to c!

This suggested to Einstein that there was something wrong with the Galilean rule for transformations between inertial frames. Using some new postulates, Einstein deduced new transformation rules between frames. The appropriate transformations are called the Lorentz transformations (LT).

In summary, the Lorentz transformations between frames are

$$t' = \gamma\left(t - \frac{ux}{c^2}\right) \tag{13.1}$$

$$x' = \gamma(x - ut) \tag{13.2}$$

$$y' = y \tag{13.3}$$

$$z' = z \,, \tag{13.4}$$

where we have introduced the conventional shorthand notation ("gamma"), $\gamma = 1/\sqrt{1 - u^2/c^2}$.

One notices a limiting speed, beyond which γ becomes imaginary, which can be identified physically with the speed of electromagnetic waves in vacuum.

The γ factor is essentially 1 for speeds $u \ll c$ and rises slowly at first and becomes appreciable only when u is a significant fraction of the speed of light. So for low speeds compared to the speed of light, the Lorentz transformations are well approximated by the Galilean transformations as we expect. For that reason most relativistic effects are not manifest in our daily encounters unless one makes a special effort to seek them.

The LT's imply the relativity of simultaneity, time intervals and lengths, all of which have been empirically verified directly or indirectly.

Velocity transformations: Suppose a moving particle is observed from the two frames of reference. The relationship between the displacement of the particle as measured from the two frames is given by

$$\Delta t' = \gamma(\Delta t - \frac{u}{c^2}\Delta x) \,, \tag{13.5}$$

$$\Delta x' = \gamma(\Delta x - u\Delta t) \,. \tag{13.6}$$

Dividing these equations gives the x-component of the particle's velocity in the S$'$ frame, $v_x' = \Delta x'/\Delta t'$, in terms of its velocity in the S frame,

$$v_x' = \frac{v_x - u}{1 - uv_x/c^2} \,. \tag{13.7}$$

This differs from the Galilean rule but reduces to it when $u/c \to 0$.

13.1.1 Relativistic Energy

Using T to denote the relativistic kinetic energy, and $\gamma = \gamma(v) = 1/\sqrt{1 - v^2/c^2}$, we have the following expressions relating the energy E of a moving particle (of speed v) to its mass m:

$$E \equiv \gamma mc^2 \tag{13.8}$$

$$T \equiv E - mc^2 = (\gamma - 1)mc^2 \quad (13.9)$$
$$E_0 \equiv E(v = 0) = mc^2 \quad (13.10)$$

The last is the famous equation which shows that mass is a form of potential energy.

Equation (13.8) indicates that for $m \neq 0$, increasing speed requires an ever-increasing amount of energy, going to infinity as the velocity tends to the speed of light. Hence particles with real mass must travel slower than the speed of light, which is thus the limiting value for energy or information transfer.

13.2 Quantum Mechanics

At the start of the 20th century, as matter was probed at small distances, it was discovered that Newtonian mechanics could not give a consistent description of phenomena such as the discrete nature of spectral lines.

It took many years before a consistent theory of microscopic matter emerged. This new theory is called Quantum Mechanics.

Thus, Newton's Laws are an approximation: If we are concerned only with the macroscopic properties of matter, then Newtonian mechanics suffices as a very good description (if, in addition, velocities are low, and gravity weak, so that relativistic corrections may be ignored).

13.2.1 Photons

Einstein, building on an earlier work of Planck, proposed that light actually propagated in discrete energy packets called photons, with the energy of each photon given by the relation

$$E = hf\,, \quad (13.11)$$

where f is the wave frequency and h is Planck's constant; $h = 6.62618 \times 10^{-34}$ Js.

That is, the phenomenon of electromagnetic waves is a classical approximation to the quantum theory of photons.

Note how the discrete particle-like nature of photons nevertheless depends on the frequency of a wave: This is already a hint of a wave-particle duality inherent in Quantum Mechanics.

13.2.2 Atoms

At the end of the nineteenth century it was thought that the atom was a microscopic version of the solar system: A positively charged nucleus with negatively charged particles moving around it. However, this leads to an inconsistency : An electron moving in a closed orbit at constant speed is actually accelerating since its velocity changes direction. Since an accelerating charge always radiates, the electron would continuously lose energy, spiralling. Hence, according to Newtonian physics, atoms are inherently unstable, a "prediction" that is contradicted by empirical facts.

In the 1900's, pioneers like Niels Bohr, de Broglie and others, inspired by the quantum postulate of Planck, developed *ad hoc* rules to explain the stability of the atom.

Although the modern perspective of atoms is different, for our purposes here it is sufficient to consider the crude Bohr Model in which the negatively charged electrons orbit a central positively charged nucleus at certain radii fixed by some quantisation rules. An electron moving in an allowed Bohr orbit does not radiate.

Each orbit can hold a certain maximum number of electrons. Atoms of different elements have different characteristic energy levels which are occupied by its electrons.

Electrons normally occupy the lowest energy states but can be excited to higher energy levels by an external input of energy. However, this excited state is short-lived and the electron returns to its stable ground state, emitting the excess energy in the form of a photon. The frequency f of the emitted radiation is related to the difference between the excited E_2 and ground state energy level E_1 according to the relation

$$E_2 - E_1 = hf \tag{13.12}$$

This model could explain the discrete spectral lines that were observed in experiments.

13.2.3 Particles and Wave

In 1923, de Broglie conjectured that a particle might have a wave associated with it, just like an electromagnetic wave had particle characteristics (the photon). For a particle of mass m moving at velocity

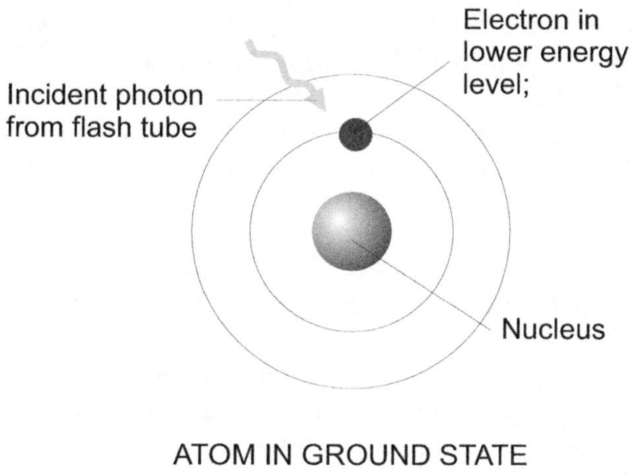

Figure 13.1: Absortion of energy by an atom

v, de Broglie postulated an associated wave with a wavelength given by

$$\lambda = \frac{h}{mv} \tag{13.13}$$

The actual meaning of the wave was only understood later: The de Broglie wave is not a physical wave, but instead it is a "probability amplitude" (see below).

13.2.4 Schrodinger's Equation

The initial framework of quantum theory, as summarised above, consisted of an *ad hoc* mixture of classical concepts and some new rules. Later, a fully consistent Quantum Mechanics was developed from which classical mechanics is recovered as an approximation valid in our macroscopic world.

Unlike classical mechanics which deals with well-defined particle trajectories obtained as solutions of Newton's Second Law, quantum mechanics deals with probability amplitudes, $\psi(x,t)$ which are solutions to Schrodinger's equation[1] $i\hbar\frac{\partial\psi}{\partial t} = H\psi$. The probability density of the particle being at a certain location x at time t is given by $|\psi|^2$.

[1] H is called the Hamiltonian. In the form we have written the equation, it is a differential operator.

Although the ordinary world we perceive by our physical senses leads to Newtonian physics, reality as described by quantum mechanics is counter-intuitive. This is not surprising since human senses have developed to cope with a macroscopic classical environment and all our mental pictures and language reflect our experience in the macroscopic domain.

13.3 Nucleus

Among the 2500 known nuclides, fewer than 300 are stable. The others are unstable. The unstable can become stable by emitting particles or electromagnetic radiation. The process by which an unstable nucleus "decays" into a more stable configuration is called nuclear decay.

There are three major types of nuclear decay: Alpha decay (by emission of Helium nuclei), beta decay (by emission of electrons or positrons) and gamma decay (by emission of electromagnetic radiation).

The equation $\frac{1}{N(t)}\frac{dN(t)}{dt} = -\lambda$ is the starting point for the law of radioactive decay (See Q18 in Chap.6). The constant λ is called the decay constant. It is a characteristic of an atom and does not depend on environmental factors such as pressure, temperature, or concentration. It is related to the mean lifetime, $\tau = 1/\lambda$.

The time taken for a radioactive sample to decay to one half of its original quantity is called its half-life, $t_{1/2} = \ln 2/\lambda$.

13.4 Fundamental Forces

There are four fundamental forces of Nature: Gravity, electromagnetism, the strong force and the weak force. We have already discussed the first two in earlier chapters.

The strong force is responsible for keeping nucleons confined inside the nucleus. It is stronger than the electromagnetic force but has a shorter range: Its effects are not manifest far beyond a nuclear diameter.

The weak force is extremely weak, and short ranged, but it is responsible for radioactive decay and so plays a very important role.

What about other forces that one encounters, like friction or "hydrogen-bonds"? At the microscopic scale, those forces are of electromagnetic origin.

Appendix A

Properties and Formulae

Exponents and Logarithms

- For $a, b > 0$,

$$a^{-1} = \frac{1}{a}. \tag{A.1}$$

$$a^0 = 1. \tag{A.2}$$

$$a^{xy} = (a^x)^y. \tag{A.3}$$

$$a^x a^y = a^{x+y}. \tag{A.4}$$

$$(ab)^x = a^x b^x. \tag{A.5}$$

$$\sqrt{ab} = \sqrt{a}\,\sqrt{b}. \tag{A.6}$$

- For $a > 1$, $y = a^x$ is positive and an increasing function of x along the real line; for $0 < a < 1$, a^x is a decreasing function of x.

- The basic relation between exponents and logarithms is

$$y = b^x \Leftrightarrow x = \log_b y\ . \tag{A.7}$$

- For $a, b > 0$, $a \neq 1$, $b \neq 1$ and $P, Q > 0$,

$$\log_b 1 = 0. \tag{A.8}$$

$$\log_b PQ = \log_b P + \log_b Q. \tag{A.9}$$

$$\log_b \frac{P}{Q} = \log_b P - \log_b Q. \tag{A.10}$$

$$\log_b P^c = c \log_b P. \tag{A.11}$$

$$\log_b P = \frac{\log_a P}{\log_a b}. \tag{A.12}$$

$$\log_b a = \frac{1}{\log_a b}. \tag{A.13}$$

- For $a > 1$ and $x > 0$, $y = \log_a x$ is an increasing function of x; it is positive for $x > 1$, negative for $x < 1$ and vanishes at $x = 1$. The graph of $y = \log_a x$ may be obtained by reflecting the exponential curve $y = a^x$ about the line $y = x$.

Binomial Expansions

For a positive integer n, the Binomial Theorem states

$$\begin{aligned}(a+b)^n &= a^n + \binom{n}{1}a^{n-1}b + \binom{n}{2}a^{n-2}b^2 + \binom{n}{3}a^{n-3}b^3 + \ldots \\ &+ \binom{n}{r}a^{n-r}b^r + \ldots + \binom{n}{n-1}ab^{n-1} + b^n \end{aligned} \tag{A.14}$$

where the binomial coefficient is defined by

$${}^nC_r \equiv \binom{n}{r} = \frac{n!}{r!(n-r)!} \tag{A.15}$$

and $n! \equiv n \times (n-1) \times (n-2) \times \ldots. \times 2 \times 1$ with $0! \equiv 1$.

Note that the $(r+1)$-th term in the above expansion is given by $\binom{n}{r}a^{n-r}b^r$.

Polynomials

- **Remainder Theorem**: If the polynomial $P(x)$ is divided by $(x - \alpha)$, the remainder is $P(\alpha)$.
  In other words, $P(x) \equiv (x - \alpha)Q(x) + R$ with $P(\alpha) = R$.

- **Factorisation Theorem**:
  $P(\alpha) = 0$ if and only if $P(x) \equiv (x - \alpha)Q(x)$.

- For the quadratic equation $ax^2 + bx + c = 0$,

$$x = \frac{-b \pm \sqrt{\Delta}}{2a} , \tag{A.16}$$

  where

$$\Delta \equiv b^2 - 4ac \tag{A.17}$$

  is called the **discriminant**. The two roots are real if and only if $\Delta \geq 0$; the case $\Delta = 0$ corresponds to a repeated root.

- The shape of the curve $y = ax^2 + bx + c$ is determined by the sign of a: When $a > 0$, the curve has a minimum point while for $a < 0$ it has a maximum. For a cubic curve $y = ax^3+bx^2+cx+d$, the sign of the leading coefficient again determines its the main shape. If $a > 0$, the curve rises upwards for large positive x and decreases for large negative x. In between, it will either have a local minimum and a local maximum or gently undulate.

Partial Fractions

- Given a rational function $P(x)/Q(x)$, its decomposition into partial fractions is as follows: First use long division to reduce the degree of the numerator to below that of the denominator. Next, each factor of $(x - a)$ in $Q(x)$ would require a partial fraction $A/(x - a)$.

- If the factor is repeated in Q, for example $(x - a)^2$, then one uses two partial fractions $A_1/(x - a)$ and $A_2/(x - a)^2$ for that factor.

- If Q contains a term that cannot be factorised (using real numbers), for example $x^2 + x + 1$, then the partial fraction for that term is of the form $(Ax+B)/(x^2+x+1)$, that is, the numerator is one degree lower than the denominator.

Simultaneous Equations

- A $m \times n$ matrix A, multiplying from the left to a $n \times p$ matrix B, yields a $m \times p$ matrix C; that is $C = AB$. The c_{ij} element of C is obtained by multiplying the i-th row of A to the j-th column of B term by term and adding the pieces.

- The pair of simultaneous equations in the variables (x, y):

$$ax + by = e \quad \text{(A.18)}$$

$$cx + dy = f\,, \quad \text{(A.19)}$$

may be written as the matrix equation $MX = R$ with

$$M = \begin{pmatrix} a & b \\ c & d \end{pmatrix}, \quad \text{(A.20)}$$

$$X = \begin{pmatrix} x \\ y \end{pmatrix} \quad \text{(A.21)}$$

and

$$R = \begin{pmatrix} e \\ f \end{pmatrix}. \quad \text{(A.22)}$$

- The **determinant** of the 2×2 matrix M is defined by det(M)$=$ $ad - bc$. If det(M)$\neq$ 0, one may define an **inverse matrix**

$$M^{-1} = \frac{1}{\det(\mathrm{M})} \begin{pmatrix} d & -b \\ -c & a \end{pmatrix} \quad \text{(A.23)}$$

and the solution to the problem is

$$X = M^{-1}R = \frac{1}{\det(\mathrm{M})} \begin{pmatrix} de - bf \\ -ce + af \end{pmatrix}. \quad \text{(A.24)}$$

- If det(M)$=$ 0 the matrix is termed **singular** and there is no inverse matrix. The inverse matrix satisfies

$$MM^{-1} = M^{-1}M = \begin{pmatrix} 1 & 0 \\ 0 & 1 \end{pmatrix}, \quad \text{(A.25)}$$

the last matrix being the **identity matrix**, usually denoted by the letter I.

Trigonometry

- For angle measurements in **radian**, $\theta \equiv s/R$ where s is the arc length of circle subtended by that angle. Therefore 2π radians equals $360°$.

- For a right-angled triangle ABC, with $C = 90°$, $\sin A = a/c$, $\cos A = b/c$ and $\tan A = \sin A/\cos A = a/b$, where the small case letters denote lengths opposite the corresponding angles.

- The sin and cos functions range over the interval $[-1, 1]$ while the tan functions ranges over the real line.

- Periodicity: $\sin(\theta + 360°) = \sin(\theta)$, $\cos(\theta + 360°) = \cos(\theta)$ and $\tan(\theta + 180°) = \tan(\theta)$.

- The **principal values** for the inverse functions $\sin^{-1}$ and $\tan^{-1}$ are those that lie in the range $-\pi/2 \le y \le \pi/2$ while the $\cos^{-1}$ function has the range $0 \le y \le \pi$.

- The functions csc, sec and cot are reciprocals of the sin, cos and tan functions ($\csc x$ is also written as cosec x).

- Phenomena that are described by an equation of the form $y(x) = A + B\sin(kx + C)$ are called **sinusoidal**.

- Some identities:

$$\sin^2 A + \cos^2 A = 1 . \tag{A.26}$$

$$1 + \tan^2 A = \sec^2 A . \tag{A.27}$$

$$\sin(-A) = -\sin A . \tag{A.28}$$

$$\cos(-A) = \cos A . \tag{A.29}$$

$$\tan(-A) = -\tan A . \tag{A.30}$$

- Addition Formulae:

$$\sin(A \pm B) = \sin A \ \cos B \pm \cos A \ \sin B . \tag{A.31}$$

$$\cos(C \pm D) = \cos C \ \cos D \mp \sin C \ \sin D . \tag{A.32}$$

$$\tan(A \pm B) = \frac{\tan A \pm \tan B}{1 \mp \tan A \ \tan B} . \tag{A.33}$$

- Factor Formulae:

$$\sin A \pm \sin B = 2\sin\frac{A\pm B}{2}\cos\frac{A\mp B}{2}\,. \tag{A.34}$$

$$\cos A + \cos B = 2\cos\frac{A+B}{2}\cos\frac{A-B}{2}\,. \tag{A.35}$$

$$\cos A - \cos B = -2\sin\frac{A+B}{2}\sin\frac{A-B}{2}\,. \tag{A.36}$$

- R-Formulae:

$$a\cos\theta \pm b\sin\theta = R\cos(\theta\mp\alpha)\,. \tag{A.37}$$

$$a\sin\theta \pm b\cos\theta = R\sin(\theta\pm\alpha)\,. \tag{A.38}$$

with $R = \sqrt{a^2+b^2}$, $\tan\alpha = b/a$.

Properties of Triangles

- The area of a triangle ABC, with small case letters denoting lengths opposite the corresponding angles, is

$$\text{Area} = \frac{1}{2}ab\sin C\,. \tag{A.39}$$

- The **sine rule**:

$$\frac{c}{\sin C} = \frac{b}{\sin B} = \frac{a}{\sin A}\,. \tag{A.40}$$

Each ratio in the formula equals the diameter of a circle which circumscribes the triangle (the centre of the circle being at the intersection of the perpendicular bisectors of the three sides).

- The **cosine rule**:

$$c^2 = a^2 + b^2 - 2ab\,\cos C\,. \tag{A.41}$$

Elevation, Depression and Bearing

- If a point P is above the horizontal through the observation point O, then the angle that OP makes with the horizontal is the **angle of elevation** of P. Similarly if P were below the horizontal then the corresponding angle would be termed **angle of depression**.

- A **bearing** denotes a direction relative to North. It is usually expressed by a clockwise angle measured in degrees. For example, 065° would refer to the direction 65° clockwise from North. Such bearings are called "absolute bearings".

- It is also convenient to use "relative bearings", which define an angle relative to a chosen axis, for example the axis along which an aircraft is pointing.

Coordinate Geometry

- If (x, y) and (x_1, y_1) are two points on the line,

$$\frac{y - y_1}{x - x_1} = m = \tan\alpha\,, \tag{A.42}$$

where $0 \leq \alpha < \pi$ is the angle the line makes with the positive x-axis. The equation for a straight line may be written in the form

$$y = mx + c \tag{A.43}$$

where m is the slope and c the intercept on the y-axis.

- If another line is perpendicular to (A.43), its slope must be $-1/m$.

- The mid-point of a line joining two points (x_1, y_1) and (x_2, y_2) is given by $\left(\frac{x_1 + x_2}{2}, \frac{y_1 + y_2}{2}\right)$.

- The equation for a circle is

$$(x - x_0)^2 + (y - y_0)^2 = r^2\,, \tag{A.44}$$

where (x_0, y_0) is the centre and r the radius. Given the form

$$x^2 + y^2 - 2ax - 2by + c = 0\,, \tag{A.45}$$

for some constants a, b, c, you can complete the squares to get $(x-a)^2+(y-b)^2 = a^2+b^2-c$; if $c < a^2+b^2$ then (A.45) represents a circle with centre (a, b) and radius $R = \sqrt{a^2 + b^2 - c}$.

- A tangent to a circle at any point P is perpendicular to the line OP where O is the centre of the circle. Therefore the **normal** to the circle at P lies along OP.

- Given three points $A(x_1, y_1)$, $B(x_2, y_2)$ and $C(x_3, y_3)$, with their relative order being anti-clockwise, the area of the triangle ABC may be obtained from the formula

$$A = \frac{1}{2}\begin{vmatrix} x_1 & x_2 & x_3 & x_1 \\ y_1 & y_2 & y_3 & y_1 \end{vmatrix} \tag{A.46}$$

where the terms are generated as follows. First, start at the left of the top row and multiply each term in the top row by a term one step to the right in the bottom row, adding the pieces: $x_1y_2 + x_2y_3 + x_3y_1$. Then start at the right of the top row and multiply each term in the top row by a term one step to the left in the bottom row, adding the pieces: $x_1y_3 + x_3y_2 + x_2y_1$. Finally, subtract the two contributions and include the overall 1/2 to get

$$A = \frac{1}{2}(x_1y_2 + x_2y_3 + x_3y_1 - x_1y_3 - x_3y_2 - x_2y_1)\,. \tag{A.47}$$

- The area of a polygon may be determined by dividing the polygon into triangles and using the above formula for each triangle.

Differential Calculus

- Let f and g be functions of x, and A, B, n represent constants in the following.

$$\frac{d}{dx}Ax^n = Anx^{n-1}\,. \tag{A.48}$$

$$\frac{d}{dx}e^x = e^x\,. \tag{A.49}$$

$$\frac{d\ln x}{dx} = \frac{1}{x}\,. \tag{A.50}$$

$$\frac{d}{dx}\sin x = \cos x\,. \tag{A.51}$$

$$\frac{d}{dx}\cos x = -\sin x\,. \tag{A.52}$$

$$\frac{d}{dx}\tan x = \sec^2 x\,. \qquad (A.53)$$

$$\frac{d}{dx}(f+g) = \frac{df}{dx} + \frac{dg}{dx}\,. \qquad (A.54)$$

$$\frac{d(fg)}{dx} = f\frac{dg}{dx} + g\frac{df}{dx}\,. \qquad (A.55)$$

$$\frac{d}{dx}\frac{f}{g} = \frac{gf' - fg'}{g^2}\,. \qquad (A.56)$$

$$\frac{df(g(x))}{dx} = \frac{df}{dg} \times \frac{dg}{dx}\,. \qquad (A.57)$$

$$\frac{df}{dx} = 1/\left(\frac{dx}{df}\right)\,. \qquad (A.58)$$

- The **stationary** points of $y = f(x)$ are those for which $\frac{df}{dx} = 0$. The stationary points are **local minima, local maxima or points of inflexion**. If $\frac{d^2 f}{dx^2} > (<)\ 0$ at the stationary point, it is a local minimum (maximum). Stationary points with $\frac{d^2 f}{dx^2} = 0$ can be examined by checking the sign of $f'(x)$ on both sides of the point. Local extrema need not be global extrema.

- If Δx is small but not strictly zero, we may form the approximation

$$\Delta y \approx \left(\frac{dy}{dx}\right) \times \Delta x\,. \qquad (A.59)$$

Differential Equations

- Consider the following first order linear differential equation

$$\frac{dy}{dx} + P(x)y = Q(x)\,. \qquad (A.60)$$

Its solution is given by $y = \frac{1}{\mu(x)}\int \mu(x)Q(x)dx$ where $\mu(x) = \exp\left[\int P(x)dx\right]$ is the integrating factor.

- Consider the following linear second-order differential equation
$$a(x)\frac{d^2y}{dx^2} + b(x)\frac{dy}{dx} + c(x)y = 0 \ , \qquad \text{(A.61)}$$
where a, b, c and y are functions of x. The equation is linear because for any constants A and B, $y = Ay_1 + By_2$ is a solution of the differential equation if y_1 and y_2 are solutions.

- If a, b, c are constants then the solutions may be found through the *ansatz* $y = e^{mx}$ which leads to the characteristic equation $am^2+bm+c = 0$. Let the the roots of this quadratic equation be p and q, $p = \dfrac{-b+\sqrt{b^2-4ac}}{2a}$ and $q = \dfrac{-b-\sqrt{b^2-4ac}}{2a}$. Then the general solution is $Ae^{px} + Be^{qx}$ where A, B are determined by boundary or initial conditions.[1]

- Let $y_p(x)$ be a particular solution of the inhomogeneous equation
$$a(x)\frac{d^2y}{dx^2} + b(x)\frac{dy}{dx} + c(x)y = f(x) \ , \qquad \text{(A.62)}$$
where f is a function of x. Then the general solution is given by $y = Ay_1 + By_2 + y_p$, where y_1 and y_2 are two independent solutions of the homogeneous (that is, $f = 0$) equation.

Infinite Series

- Taylor series: For a function $f(x)$ which is infinitely differentiable at $x = a$,
$$f(x) = \sum_{n=0}^{\infty} \frac{f^{(n)}(a)}{n!}(x-a)^n \ , \qquad \text{(A.63)}$$
where $f^{(n)}(a)$ is the n-th derivative of $f(x)$ evaluated at $x = a$.

- $(1+x)^n \ = \ 1 + nx + \dfrac{n(n-1)}{2!}x^2 + \dfrac{n(n-1)(n-2)}{3!}x^3 + ...,$ for $|x| < 1$.

- $e^x = \displaystyle\sum_{n=0}^{\infty} \frac{x^n}{n!}.$

---

[1]We assume here that $p \neq q$, which is the usual case. For $p = q$ the general solution is of a different form.

Integral Calculus

- **Fundamental Theorem of Calculus:**

$$\int_a^b f(x)dx = F(b) - F(a)\,, \tag{A.64}$$

where $F(x)$ is a function that satisfies

$$\frac{dF(x)}{dx} = f(x)\;. \tag{A.65}$$

Hence

$$\int_a^b \frac{dF(x)}{dx}\,dx = F(b) - F(a)\,. \tag{A.66}$$

- Indefinite integrals:

$$\int f(x)dx = F(x) + C\,, \tag{A.67}$$

where (A.65) still holds and C is a constant of integration which can be fixed once we have more information about the problem.

- In the formulae for indefinite integrals below, f and g are functions of x while a, b, n are constants; C is a constant of integration.

$$\int (a+bx)^n\,dx = \frac{(a+bx)^{n+1}}{b(n+1)} + C, \quad n \neq -1. \tag{A.68}$$

$$\int \frac{1}{a+bx}\,dx = \frac{1}{b}\ln(a+bx) + C\;. \tag{A.69}$$

$$\int \sin(a+bx)\,dx = \frac{-1}{b}\cos(a+bx) + C\;. \tag{A.70}$$

$$\int \cos(a+bx)\,dx = \frac{1}{b}\sin(a+bx) + C\;. \tag{A.71}$$

$$\int e^{(a+bx)}\,dx = \frac{1}{b}e^{(a+bx)} + C\;. \tag{A.72}$$

$$\int (f+g)\,dx = \int f\,dx + \int g\,dx\;. \tag{A.73}$$

- The following identity holds:

$$\int_a^c f(x)dx = \int_a^b f(x)dx + \int_b^c f(x)dx\,. \tag{A.74}$$

- For calculating areas between the curve $y = f(x)$ and the x-axis, note that if $f(x) < 0$ within a region $x_1 \leq x \leq x_2$, the integral in that region would give a negative value and the area is then the negative of the integral.

- One may also evaluate the area between a curve and the y-axis. In this case the integral would be $\int_{y_1}^{y_2} x\, dy$.

Vectors

- Consider a vector $\vec{a} = a_x\vec{i} + a_y\vec{j} + a_x\vec{k}$. Its magnitude is

$$|\vec{a}| \equiv \sqrt{a_x^2 + a_y^2 + a_z^2}\,. \tag{A.75}$$

- The dot product of two vectors:

$$\vec{a}\cdot\vec{b} = a_x b_x + a_y b_y + a_z b_z = |\vec{a}||\vec{b}|\cos\theta\,, \tag{A.76}$$

where θ is the angle between $\vec{a}$ and $\vec{b}$. Clearly $\vec{a}\cdot\vec{b} = \vec{b}\cdot\vec{a}$.

- The cross product of two vectors:

$$\vec{a}\times\vec{b} = |\vec{a}||\vec{b}|\sin\theta\;\vec{n}\,, \tag{A.77}$$

where $\vec{n}$ is the unit vector perpendicular to the plane defined by $\vec{a}$ and $\vec{b}$; its direction given by the right-hand rule (point your straight fingers in the direction of $\vec{a}$ and close them in the direction of $\vec{b}$, the thumb points along $\vec{n}$). The cross-product is anti-commutative, $\vec{a}\times\vec{b} = -\vec{b}\times\vec{a}$. In Cartesian coordinates,

$$\vec{a}\times\vec{b} = (a_y b_z - a_z b_y)\vec{i} + (a_z b_x - a_x b_z)\vec{j} + (a_x b_y - a_y b_x)\vec{k}\,. \tag{A.78}$$

- With a third vector $\vec{c}$, we have

$$\left(\vec{a}\times\vec{b}\right)\cdot\vec{c} = \vec{a}\cdot\left(\vec{b}\times\vec{c}\right)\,. \tag{A.79}$$

Complex Numbers

- Any complex number z may be written in the form $z = a + ib$ where a, b are real and $i = \sqrt{-1}$. The conjugate is $\bar{z} = z^* = a - ib$. We have

$$z\bar{z} = a^2 + b^2 \equiv |z|^2 \ . \tag{A.80}$$

- The polar representation is

$$z = Re^{i\theta} \ , \tag{A.81}$$

  where $R = |z|$ and $\theta = \arctan(b/a)$.

- The Euler identity $e^{i\theta} = \cos\theta + i\sin\theta$ relates the exponential function to the trigonometric functions.

Appendix B

Additional Resources

Here are some complementary print and web resources:

1. How Stuff Works at www.howstuffworks.com
2. HyperPhysics at
   http://hyperphysics.phy-astr.gsu.edu/hbase/hph.html
3. Wolfram Alpha (Computational Knowledge Engine) at
   http://www.wolframalpha.com/
4. Guesstimation, by L. Weinstein and J. Adams
5. A First Course in Mathematical Modelling, by
   F. Giordano, W. Fox, S. Horton and M. Weir
6. Integrated Mathematics for Explorers, by
   Adeline Ng and R. Parwani

Appendix C

Applications Index

This index points to some of the exercises, mainly in part one.

D

E

F

G

H

I

K

L

M

N

O

P

Q

R

S

T

V

W

Y

The Authors

Dr. Wei Khim Ng
is a theoretical physicist at the Department of Physics, National University of Singapore. He has taught a range of modern physics and general education modules, some of which were open to high school students. His research interests are in quantum theory, neutrino physics, nuclear physics education and astrophysics education. He is an elected life member of the Institute of Physics, Singapore and Sigma Pi Sigma, USA.

Dr. Rajesh R. Parwani
is a theoretical physicist whose research specialities are quantum theory and cosmology. He has taught modules in quantitative reasoning, multi-disciplinary science and physics. In 2013 he co-authored the book "Integrated Mathematics for Explorers". He provides research and education consultancy through the Simplicity Research Institute, Singapore.

Contact us: enquiry@simplicitysg.net

(Updated Feb 2016)
Real World Mathematics
Companion website: www.simplicitysg.net/books
Facebook: www.fb.com/realworldmath

Made in the USA
Monee, IL
04 November 2026

67381346R00085